T0292059

THE ACCELERATION OF CULTURAL CHANGE

SIMPLICITY: DESIGN, TECHNOLOGY, BUSINESS, LIFE
John Maeda, editor

The Laws of Simplicity, John Maeda, 2006

The Plenitude: Creativity, Innovation, and Making Stuff, Rich Gold, 2007

Simulation and Its Discontents, Sherry Turkle, 2009

Redesigning Leadership, John Maeda, 2011

I'll Have What She's Having, Alex Bentley, Mark Earls, and Michael J. O'Brien, 2011

The Storm of Creativity, Kyna Leski, 2015

The Acceleration of Cultural Change: From Ancestors to Algorithms, R. Alexander Bentley and Michael J. O'Brien, 2017

THE ACCELERATION OF CULTURAL CHANGE

From Ancestors to Algorithms

R. ALEXANDER BENTLEY AND MICHAEL J. O'BRIEN
FOREWORD BY JOHN MAEDA

The MIT Press
Cambridge, Massachusetts
London, England

This book was set in Scala Sans and Scala by Toppan Best-set Premedia Limited.

Library of Congress Cataloging-in-Publication Data

Names: Bentley, R. Alexander, 1970- author. | O'Brien, Michael J.
  (Michael John), 1950- author.
Title: The acceleration of cultural change : from ancestors to algorithms /
  R. Alexander Bentley ; foreword by John Maeda.
Description: Cambridge, MA : MIT Press, [2017] | Series: Simplicity:
  Design, technology, business, life | Includes bibliographical references
  and index.
Identifiers: LCCN 2017006483 | ISBN 9780262036955 (hardcover : alk.
  paper), 9780262551977 (paperback)
Subjects: LCSH: Social evolution.
Classification: LCC HM626 .B473 2017 | DDC 303.4--dc23 LC record
available at https://lccn.loc.gov/2017006483

CONTENTS

CONTENTS

FOREWORD

John Maeda

I'm pleased to present what could be easily thought of as a sequel to *I'll Have What She's Having* because the two books share authors. But with what I've learned from *The Acceleration of Cultural Change* I now know that thinking of the two books as related to each other— just because one has followed the other—is an antiquated way of thinking. For a lot can happen in the six years that have followed their first book, and especially when those years are counted by the standards of Moore's Law.

More than ten years have passed since the writing of the *Laws of Simplicity*—the iPhone had not yet been launched when the inaugural book of the Simplicity series had been published in 2006. The number of computer users didn't number in the billions, and mobile computing meant a laptop that weighed too many pounds and ran for only a few hours on a single charge. Today we live untethered, always-on, and constantly plugged-in to a myriad of machines and other human beings with what Bentley and O'Brien dub as the modern Acheulean hand ax: our always ready and available smartphone. Whereas the hand ax would be used to cut, the smartphone is used to *connect*.

Our newfound ability to use computing to connect at scales previously unknown to humankind, is what Bentley and O'Brien's new book attempts to address by looking backward to enable us to see forward. They look way, way backward—as far as 1.7 million years ago to the Pleistocene-age. But you already knew this because I figured you've already looked up "Acheulean hand ax," thinking you might be able to go out and buy it online at the Patagonia store. Sorry. There are so many terms and words in this book that aren't coming off of a trending TechCrunch article or meme-ified in the latest futuristic TED talk. It's about a lot of old stuff that no longer has relevance in this new world today where knowledge has moved from being thin and deep in its "traditional" form, to now broad and shallow—what they term as being "shaped like a horizon."

Staring out at the horizon is something that I know fairly well. I've been reaching for it to try and touch it and understand it. Although I had a front-row seat to the rise of computing within the MIT world during the '80s and '90s, I could see it leaving the realm of research and the academy, and sensing what Bentley and O'Brien have done in this book—go backward in history to reorient toward the future—similarly, I left MIT to go backward in time to run a classically minded art and design university. And then I went forward again into the future, by heading to Silicon Valley to work in the venture-capital industry as a partner at Kleiner Perkins Caufield & Byers—the investment firm that gave birth to companies we know today such as Google and amazon.com. After working with more than a hundred tech startups at all phases of growth, I decided to join one myself at the end of 2016 and now find myself at the tech startup Automattic, which was founded by the cofounder of the WordPress project.

So the approach that Bentley and O'Brien have taken is one that makes complete sense to me, when considering how we got to where we are is the pseudo-sum total of so many technological, social, economic, and political changes that are now all impacted by Moore's Law. So we need to go broad, really broad, in order to collect enough disparate data points with which to take a good enough stab at figuring out the future. What you will find in this slim, useful volume is so many dots of information scattered through time and space, and across a diverse set of cultures. And, yes, techie culture is a large ingredient in the authors' analyses, so you'll have a healthy dosage of knowledge dots that are familiar in 2017 such as the Internet of Things, machine learning, and, of course, Instagram and Snapchat.

In closing, I want you to know that only by my living in Silicon Valley did I come to realize how pervasive the philosophy of Apple cofounder Steve Jobs lives at the heart of tech. It's clear that his approach to life and the pursuit of knowledge was not unlike the one that the authors have taken in this book, as exemplified by Jobs's famous commencement speech:

Again, you can't connect the dots looking forward; you can only connect them looking backward. So you have to trust that the dots will somehow connect in your future. You have to trust in something—your gut, destiny, life, karma, whatever. This approach has never let me down, and it has made all the difference in my life.

—Steve Jobs, from his commencement speech at Stanford University,2005 (http://news.stanford.edu/2005/06/14/jobs-061505)

Look to the horizon. I leave your Bayesian mind to process the dots that Bentley and O'Brien have curated for us. I'll be looking to the horizon together with you.

Global Head of Computational Design and Inclusion
Automattic, Inc.

In the late 1980s, Alex—we'll refer to ourselves as Alex and Mike—worked at the Middleton Twentieth Century Theater in Madison, Wisconsin. This was a 1940s' corrugated-iron Quonset hut that was built in less than a week. Tickets were 99¢ for all seats, all shows, all times, and all ages. The movies shown were six months old and the ones almost no one still wanted to see. There was one screen, a monophonic speaker above it, and the skeleton of a rat in the basement that the manager would show to new employees.

Alex worked both the ticket booth and concession counter, and his initial responsibility was to take a dollar bill from the customer, and jam a penny and the sweaty ticket stub into the customer's hand. Alex would then step back out of the booth and into the lobby to sell soda, popcorn, and Mike and Ikes to the same customer(s). In the summer, he would drag out the Toro lawnmower from a closet beneath the projection booth and go outside and mow the strip of grass behind the gravel parking lot. The manager would walk out to watch as Alex sweated in a white shirt and tie, inside a cloud of brown grass bits and flying pebbles.

Change was slow to come to the Middleton. The manager showed Alex how to take inventory, which meant counting the dusty Hershey Good & Plenty boxes in the display case and subtracting the number from yesterday's total to get the sales for the day—usually a box or two. Alex earned $3.60 an hour—a couple cents more than the minimum wage at the time. The theater's air-conditioning had broken down in the early 1980s and still needed to be fixed. One night, the manager said that if no one showed up for the 9:25 show, Alex could close up early. Two people showed up, unfortunately, and bought popcorn. At the end of the film, Alex put the gross sales for the day of just over $11 in a lockable canvas bag and transported it to the outside drop box at one of the downtown banks.

Despite being only a generation ago, this scene is unrecognizable now. Most children in the United States watch movies on a

device, and cash is used in only about 10 percent of US transactions. Alex's work experience—like repairing film with a splicing device or informing customers, individually, of showing times when they phoned in—would not count for much on a modern résumé.

Thinking of movies makes us think of ratings, which were around back then, but nothing like we have now. You can now find ratings of everything imaginable—hotels, restaurants, roads, dating services, and even massage parlors. Customers are incredibly picky compared to the Middleton Theater customers, who would watch anything. Think of the modern customer's one-star motel review on TripAdvisor, complete with a half-dozen pictures of a paint chip on a dresser drawer. It makes you want to ask, "What did you expect for $39?" Compare that customer to two customers on the hot sidewalk outside the Middleton Theater in August 1990, when the manager was trying to talk them out of the movie because he had just finished sealing a pack of fifty $1 bills and they had a twenty. The manager first tried telling the couple that the movie had started ten minutes ago and that the beginning was crucial. When the couple said they didn't mind—"Two tickets, please"—the manager tried, "It's really hot in there" and told them to wait while he went to check. He returned ten minutes later with a thermometer/ humidity meter saying, "It's *really* hot in there: 85 degrees with over 90 percent humidity!" Finally, the man said, "The hell with this!" and he and his girlfriend left (the next customer, who came shortly after that, paid with a dollar bill and went straight in).

We all think we know why this 1980s' scene seems so long ago. From e-mail in the 1990s, to iPhones and Facebook in the 2000s, to a proliferation of ephemeral social media, rapid change has become part of our expectation, not just in generational gaps, but in intragenerational ones as well. Some people expect to see a time when brains are wired directly into the Internet. If that comes to pass, it certainly will be different to be human. But this book is not about what it might be like to have a chip in your brain or transhumanism more generally. In fact, this book is not about you as an individual at all. Rather, it's about your culture. More to the point, it's about those scores or hundreds of generations of people of past centuries in your cultural ancestry who passed on and contributed to the habits and knowledge that seem so normal to you now. This book is also about how that system of cultural inheritance is radically changing. It's about how the scene at the Middleton Theater represents a *process* of culture that, invisibly, differs far more than simply in terms of gadgetry.

In his 1976 book *The Selfish Gene*—long before he became a prolific and cranky Twitter user—Richard Dawkins coined the term *meme* for an idea, style, or behavior that spreads from person to person within a culture. In the mid-1990s, philosopher Daniel Dennett described a *meme's-eye view* that modeled ideas as viruses whose survival depends on their spreading among their human hosts. Applying Dawkins's criteria, memes spread through longevity, fidelity, and fecundity. In other words, successful memes are retained in the memory, and then get copied accurately and often. The Internet is the perfect medium for memes, and people talk regularly about them, especially in relation to online text, tweets, pictures, and so on, which are copied and shared. A picture of Michael Jordan crying has become so widespread that young people today might know him better by this meme than by his basketball career.

This book is not about how to spread your memes. If you want that, just read a marketing blog such as knowyourmeme.com. This book is about how culture evolves, and frankly, how culture evolution was never only about spreading memes. Evolution is about three things and three things only: variation, transmission, and sorting. Everything we discuss in this book boils down to these three components of the process that has shaped humans into the large-brained, hairless apes we are today. That's the genetic part of human evolution, but it's the cultural part that has shaped and continues to shape what humans everywhere do and say.

Thinking of the new world in terms of memes and your individual experience on a smartphone is fun, but it doesn't really get us anywhere. We are going to ask you to think a bit harder about things. We're going to draw on quite a few different sciences, from anthropology to archaeology, economics, evolutionary biology, and even (briefly) physics. Most important, we will ask you to think on a different scale than you might be used to—one that includes many, many people, over many generations, sharing and tweaking different units of culture.

Humans have evolved to learn cultural know-how and teach it to the next generation, with occasional small adjustments that can track environmental change. The knowledge embedded in culture has instructed people in how to address environmental challenges, feed themselves and their group, and efficiently store that knowledge as cultural practice, which makes it learnable and heritable. Humans became cultural animals through individual characteristics such as large brains and long life spans, but also through group-level features such as kinship networks and specialization of knowledge.

These features, however, often stand in stark contrast to those that define us today. Fewer people now inherit their occupations; technological change has become so rapid that previous generations of knowledge are seen as irrelevant; we are no longer learning from the wisest individuals in the group; and the new, online world is filled with masquerading "experts," both human and nonhuman. How do we plan for this world? How will knowledge accumulate if learning occurs through different pathways than it has for the past hundred thousand years? How will knowledge be sorted? Through a survey of some of the main technologies that are reconfiguring the way we learn, this book explores the implications for the future of cultural evolution.

Our central premise is that the shape of cultural transmission has changed dramatically over recent decades, from one that is thin and deep to one that is shallow and broad. Thin and deep—what we can call "traditional"—is the shape of local learning of knowledge that is inherited through ancestors and changes slowly, on the time scale of many generations. Traditional knowledge is finely tuned to the local environment over many generations of slow cultural adaptation. Broad and shallow—shaped like a "horizon"—describes recent knowledge, or just information, that is shared widely, including potentially internationally. In the horizon regime, the tempo of knowledge creation has accelerated to the point of little connection with ancestral knowledge.

This metaphor, from thin and deep to shallow and broad, underlies this book and divides it into two parts. The first five chapters are about traditional aspects of cultural evolution. The transition occurs in chapter 6, where we look at how certain long-term traditions, such as marriage and diet, are rapidly changing through shallow and broad horizons. Our shallow and broad discussion continues in the next few chapters through the science of networks, prediction markets, and the explosion of digital information. Finally, in chapter 10, we ask whether artificial intelligence may solve this overload problem by learning to integrate concepts over the vast idea space of digitally stored information through time. Although we are not going to wax poetically about a "singularity," it could be unprecedented in the hundreds of thousands of years of human culture.

We take this opportunity to thank Bob Prior, executive editor of the MIT Press, for his unflagging support of the project. We also thank John Maeda, editor of the Simplicity: Design, Technology, Business, Life series, published by the MIT Press, for graciously accepting our book into his series. This is the second book we have published with Bob and John—the other being *I'll Have What She's Having: Mapping Social Behavior* (2011). Finally, we thank Gloria O'Brien and the MIT Press's Deborah Cantor-Adams for providing excellent editorial suggestions.

TRADITIONAL MINDS

The janitor at the Middleton Theater, Sam, was in his thirties and drove a Firebird with the window open in any weather. Consequently, his long black hair always looked feathered. Sam would come in for work in the afternoon, before the bars opened, wearing white pants, silk shirt, and sunglasses, disappear into the theater for a while, and then emerge without a spot. He spent most of that time searching the place for wallets and loose change, tossing out bins of popcorn as he went. When he was done, Sam would put himself down on the time sheet for about five hours of work. That's because Sam spent most of that time chatting to people working there, meaning Alex and the manager, about all sorts of things, including the Shroud of Turin, in which he had considerable interest. Not a lot of knowledge but a lot of interest.

Whether telling stories, singing songs, or texting, humans have always liked to chat with each other at work. Traditionally, humans are viewed as belonging to a unique species because we are large-brained, bipedal, and highly social—not just a smart species that can also be social, but a species that *evolved* to be social. Neuroscience is confirming the essential social function of the brain: functional magnetic resonance imaging scans show that social exclusion, bereavement, and being treated unfairly can activate the pain network of the brain. Conversely, having a good reputation, being treated fairly, cooperating, giving to charity, and even schadenfreude (getting pleasure from another's misfortune) can all activate the brain's reward network.

Large human brains evolved to be social, but why? The evolutionary costs of a large brain are considerable. According to a twentieth-century theory called the *obstetrical dilemma*, for example, a large cranium increases the risk to a mother giving birth, which is part of the reason hundreds of thousands of women die each year from pregnancy or childbirth-related causes. In terms of evolution, a large risk must be justified by some comparably large benefit. The benefit that arguably most makes up for these costs is social cooperation, because groups can survive better through the internal cooperation of individuals. Where this cooperation is lacking, due to a lack of health care and/or poverty, maternal death rates are higher. Prehistorically, what distinguished humans from primates was the social context of childbirth, especially midwifery. Many causes of maternal death, such as hemorrhage, were avoided through "folk" medicine. Goddesses from the Aztecs to the ancient Egyptians were depicted as powerful women giving birth, in upright positions, while assisted by other women.

The earliest form of cooperation among our ancestors 1.5 million years ago likely was in terms of hunting for and sharing meat. More meat in the diet was essential to the evolution of the large human brain, as Leslie Aiello and Peter Wheeler contended. Because humans could get calories more easily, they evolved larger brains and smaller guts. Once our early ancestors could control fire, Richard Wrangham observed, they could cook food and afford to have a smaller gut because cooked starches are much easier to digest.

Being social also means competing socially, which requires brain power. Primates "waste" a great deal of time grooming each other because the close communication helps them get "gossip" about potential mates, according to anthropologist Robin Dunbar, thus helping them reproduce those social genes. The larger the group, the more cognitively demanding this becomes, and Dunbar argued that human language evolved in place of grooming. In the early 1990s, Aiello and Dunbar famously compared the average group size among primates to brain size—technically the size of the neocortex—and found a clear relationship. The correlation was slightly different for monkeys than for apes, but in both, the larger the neocortex, the larger their typical social group. Back then, Aiello and Dunbar's aim was to extrapolate this curve so that they could estimate the typical group size of our hominid ancestors, such as *Homo habilis* and *Homo erectus*, using the volume of the brain that can be measured from fossil remains.

Modern humans have a brain size of about 1,400 cubic centimeters, and this is the number that Dunbar used to arrive at his estimate of 150 as being the typical limit of real—defined as *meaningful*—social relationships that a person will have. Dunbar's number has proven so prophetic that a 2015 study of US teenagers showed that a typical Facebook user has 145 friends and a typical Instagram user about 150 followers. Think about this for a minute. In the early 1990s, before most of us had even heard of the Internet, an anthropologist compared the brain size of different species of apes to their typical group sizes observed in the wild. He then extrapolated the correlation out to the brain size of humans. Twenty-five years later, this extrapolation predicts the typical number of Instagram followers among US teenagers. This is pretty amazing.

Social learning has become the focus of study for behavioral scientists from a range of fields, including psychology, anthropology, and economics. Our success as a species, wrote economist Samuel Bowles, relies on the correct social and networking skills of knowing who, what, and when to copy. Whereas earlier research on conformity focused on adults, recent psychology experiments demonstrate conformity among children or even infants, who will learn from adults whom others are seen to be looking at and learning from. Other kinds of learning are biased toward natural categories, such as plants. Whereas infants will put most any plastic toy in their mouths, they will hesitate when given a plant, watching first for cues from an adult as to whether the plant is edible or poisonous, and then proceeding accordingly.

The fact that many human cultures share in the parenting of children underlies anthropologist Sarah Hrdy's belief that sharing and cooperation, more than competition, is what makes us human. Some years ago, Alex and his family stopped at a family-owned burrito place in San Bernardino, California, and after the order was placed, the woman behind the counter asked, "Can we hold your baby?" Without much hesitation, Alex and his wife handed him over, and the woman cuddled him for several minutes while a couple of the other staff came over, and then handed him back, along with the burritos.

This simple scene would be impossible to duplicate with any other primate, as all other primate mothers would fight pretty much to the death if you tried to take their infants away. A wild ape mother will not let others hold or carry her baby, wrote Hrdy, adding that the only other primates who share infant care are marmosets and tamarins. Helping in other ways with child-rearing, however, can be seen among macaques, squirrel monkeys, meerkats, and scrub jays. This improves infant survival and later reproduction. Before all the social structures that affect human child-rearing— family structure, wealth inheritance, religion, and so on—what made humans unique, said Hrdy and others, was the sharing of food, long life of females after menopause (so grandmothers can help their daughters raise children), and fact that human infants could form attachments with multiple caregivers. All this requires a social brain, alongside emotion, empathy, and a theory of mind. In Hrdy's terms, humans are evolved to be cooperative breeders, which means that the isolation of women in the home is not a helpful state.

The act of sharing food, such as tubers dug from the ground or the meat brought back from a hunt, is probably as old as our species or even genus. The Latin origin of *companion* is "one who breaks bread with another." To solidify relationships between families, the Bantu of Zimbabwe would exchange food in a "clanship of porridge," for example. At family dinner tables across the globe, a child in a snit can often be brought back into the conversation with an offer of more food, especially dessert.

It is no surprise, then, that sharing is what so many people do in their electronic relationships. Well ahead of the curve, anthropologists Heather Horst and Daniel Miller recorded aspects of cell phone usage in Jamaica in the early 2000s. Their research was a combination of listening, observing, and interviewing, and then collecting information about the contacts individuals had on their phones. This was back in a setting where a mobile phone was mainly a device for calling or texting friends and family. Horst and Miller found that family and kin were most important to women, who led relatively sheltered lives and, on average, had fewer than thirty numbers saved in their phones. House phones were great for long, deep, and protracted relationships, but not so well suited to maintaining connections over time with acquaintances, as Horst and Miller showed, quoting one woman who hadn't called her male friend for a while: "Him ask if mi get rich an switch, that's what he call mi an' ask mi," asking whether she was too good for him now.

PRODUCERS AND SCROUNGERS

Computer simulations allow scientists to consider many interactions and large numbers of agents at the same time. Back in 2010, St Andrews University's Kevin Laland and his colleagues hosted a computer-algorithm tournament that was rooted in Robert Axelrod's notable 1984 computer contest that pitted different repeated prisoner's dilemma algorithms against each other. Laland's "Social Learning Strategies Tournament" was built around proposals for the most successful default strategy for an agent in a large social environment. The contest entries consisted of software code that instructed an agent on how to interact with other agents it met over a number of rounds. The tournament creators expected the winner would have a superior social-learning strategy about whom and when to copy. Mere random copying was not seen as likely to win because information can be wrong as well as outdated.

The entries reflected the broad interest in the topic, with biologists, anthropologists, psychologists, economists, and mathematicians all jumping in. The winners, as much a surprise to themselves as to the expert panel overseeing the tournament, were two Canadian postgraduate students, Dan Cownden (a neuroscientist) and Tim Lillicrap (a mathematician), neither of whom were social-learning experts. They labeled their entry "discount machine." Its basic instruction was to copy—and copy often—and bias that copying in favor of recent successful strategies, thus "discounting" older information. It was not quite random copying, but close: copy any success, just as long as it is a *recent* success.

Similar games shed light on how the success of a social-learning strategy depends on what other strategies are being played out in a group. Mike and his colleague Alex Mesoudi created an experiment where the participants played a computer game in which they "made" stone projectile points designed to hunt bison. The participants were allowed to change aspects of the stone points—for instance, the length and width—and then see how well their points would perform (based on archaeological knowledge) on bison hunts. After each round, the hunters could see their own scores (in caloric return) compared to the scores of the other hunters, and they could also see the different designs that others were using. Each hunter could invent new shapes or copy others whose hunting success scores they could see. In all runs of the game, social learners scored better than those who refused to copy the success of others.

But it really is a bit more complicated because people tend to work in groups, and if everyone always copies someone else—meaning there is no innovation—then the group runs the risk of extinction. Mike and Alex were able to see what the proportion was of copying and inventing within the entire *group* of projectile point makers. They then charted the group's success as a function of this ratio and found that there seemed to be an optimal mix of new information created by the minority "producers" and the majority of people who copied them—"scroungers." We would expect that over time, or in a well-governed community, a nice mix would settle out, with some producers and a majority of discerning scroungers.

What is a "nice mix," though? It turns out that a number of studies have shown that somewhere around 5 percent producers is ideal. The Max Planck Institute's Ian Couzin and his colleagues, for example, showed that within a flock of birds, it takes only a small fraction of individual learners (producers) to impart a coherent direction to the entire group, as the majority copies the travel direction of its neighbors. Individual thinking and accurate information among these few producers are critical, because false alarms can spread and amplify across a flock through bad information. Couzin and his colleagues demonstrated how ignorance or ambivalence among the majority allows the consensus to be controlled by a small but determined minority. There is a rapid transition from the ambivalent majority rule to the determined minority rule as either the minority is made more determined or ignorance and/or ambivalence among the majority is increased.

Scroungers can proliferate when there isn't a lot of change in the environment, especially when it is costly or risky to produce new ideas, but having too many scroungers compared to producers leaves you with a noisy echo chamber. As Mesoudi pointed out, when copiers become too numerous, they will be copying from each other so much that the quality of their information deteriorates in the echo chamber. We can think of Internet availability as raising the number of information producers not to 5 percent of the community but rather to 5 percent of the world. The obsession with novelty and individualism is not only a disadvantage; it is WEIRD, as in the acronym for "Western, educated, industrialized, rich, and democratic" societies, coined by Joe Henrich and his colleagues. When Mesoudi and his colleagues took the projectile point experiment to a non-WEIRD community in a provincial Chinese city, they found that the inhabitants tended to copy more than did the British nationals, Chinese immigrants in Britain, or residents of Hong Kong. As a result, the Chinese scored better with their projectile points, both in groups and individually. Westerners tended to persist in their individual learning, and their scores suffered for it. Differences in cultural individualism may take centuries to develop. The history of rice farming in southern China, for instance, underlies a more interdependent and holistic culture compared to the wheat-growing populations of northern China, according to a psychological study led by Thomas Talhelm of the University of Chicago.

CULTURAL INTELLIGENCE

Evolutionary anthropologists have argued that we are wrapped in culture, or addicted to culture, in that we inherit our instructions for how to cope in the world from past generations. Under this cultural-intelligence hypothesis, the cumulative cultural toolbox contains a different set of skills than just individual intelligence and problem solving. Humans are evolved not to solve problems on their own, or even solve them cooperatively on the fly, but rather to accumulate knowledge collectively, over generations, and apply that knowledge to the same environment (roughly) in which previous generations lived. The environment, however, has two components—cultural as well as physical—and as we will see throughout the book, humans are now living in cultural environments that are radically different from those of their ancestors—even recent ancestors. The fast pace of cultural change is having enormous impacts on human evolution.

Ironically, no matter the pace at which it changes, culture still provides the basis for humans to exist. It has to. We are so far into the game that without culture—literally, the high capacity for cooperation and learning—humans would be lost. Consider hunter-gatherers, who live in small, highly mobile bands and exploit wild resources over a wide area. Although both chimpanzees and hunter-gatherers share food—for example, meat—only humans exchange nonfood items as gifts to maintain social relationships between groups. Because their population density is low, this exchange has benefits. In the Kalahari Desert, a Ju/'hoansi band could use the water holes of their exchange partners when theirs dried up in a drought. Until the mid-twentieth century, the Ache hunter-gatherers of eastern Paraguay hunted mammals with bows and arrows and foraged for plant foods. A regional group of Ache was something over 500 people living in twenty or so residential bands at least ten kilometers apart.

In Tanzania, the eastern Hadza still hunt and gather with bows, axes, and digging sticks. About a thousand of them live in some fifty bands, some less than a kilometer apart, and others as much as eighty kilometers apart. Anthropologist Kim Hill and his colleagues found that the average Ache or Hadza man will have interacted with about 300 to 400 different men in his lifetime. When you roughly triple this, to account for opposite-sex adults and children, the number of interactions is much higher than for chimpanzees, which is about 20, and much higher than Dunbar's number of about 150. Remember, though, that Dunbar was referring to meaningful relationships, not the one-off kind.

We see this with the Hadza. When a group led by Coren Apicella asked more than 200 women and men, from seventeen distinct camps, to whom they would give a gift of honey, they found the average number of gift partners was about six. Despite the apparent messiness of the network, which makes it appear as if everyone is connecting with everyone else, they're really not. Despite everyone knowing everyone else, meaningful relationships involve a much smaller number of people.

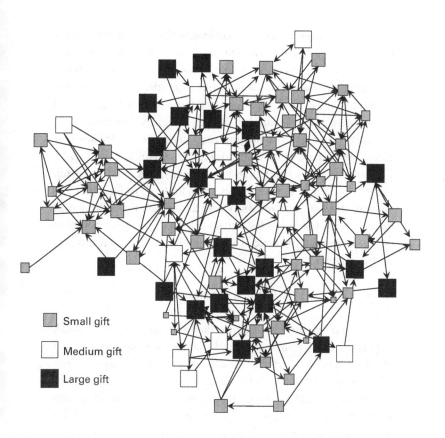

Small gift

Medium gift

Large gift

Now compare this to a Christmas card study that Dunbar did with his student Russell Hill in England in the days before social media, which found the average number of cards sent was close to 150—the Dunbar number. Not surprisingly, Hill and Dunbar also discovered that self-rated emotional closeness decreased between pairs of friends as the frequency of contact decreased. So we may say "keep your friends close but your enemies closer," yet in reality we want to keep our best friends closest.

But do we know who those people are? Can we reliably identify our "close friends"? Hill and Dunbar's study is interesting, but it was built on the self-rated closeness someone feels for someone else. In other words, you identify your close friends and then rank them in terms of closeness. A recent study by Alex Pentland and his group at MIT's Media Lab, however, showed how people are not particularly good at judging their asymmetrical friendships. Like King Richard II of England and his cousin Henry IV (who imprisoned him in 1399), each would clearly rank their friendship differently. This lack of reciprocity could lead to hurt feelings—you suddenly find out that you feel closer to a person than that person does to you—but more than that, it limits a person's ability to engage in mutually beneficial arrangements. To help sort this out, we have names.

WHAT'S IN A NAME?

Maybe their importance is diminished in today's Western world, but names have traditionally played an important role in society and still do today in much of the world. Names can inform us about our inherited biological and social relationships, which were crucial for a social species that became organized by kinship systems that controlled not just marriages but also alliances, food sharing, wealth inheritance, and the specialized jobs and crafts that people engaged in that contributed to the overall success of a group. On a sparsely populated landscape, where exchange with kin is essential for survival, knowing how to interact socially with those with whom you have only occasional contact is crucial. This starts with names and how to address someone.

Hunter-gatherer societies tend to be small, but their kinship-naming systems can be complex and highly informative. In the early 1950s, Lorna Marshall of Harvard's Peabody Museum traveled to southwestern Africa's Kalahari Desert to study the !Kung Bushmen (now Ju/'hoansi). During a fourteen-month stay, Marshall was able to record a kinship network of about six hundred people. They did not know the proximate reasons for their kinship system looking the way it did, but they did have a keen sense of cultural transmission. The !Kung would say, "God created people and told them what terms to use for each other. Parents have taught their children ever since what terms to use." Marshall then asked people what kinship terms they used for each other, and while she found the terms were straightforward for parents, children, and siblings, they were less so for grandparents or grandchildren, where one of two terms might be used—one for biological relatedness and the other as a way of classifying individuals with the same names.

The kin terms were intricate and precise. Terminologically, no two people were simultaneously interrelated in more than one way, even though their names may have changed on marriage. The names conveyed information about the relative biological age— nothing surprising here, as our terms "mother" and "daughter" do the same thing. There were five parent-child terms, which were never modified or used for any other purpose, and three sibling terms for elder brother, elder sister, and younger brother or sister that were never modified but occasionally were used for another purpose. On top of this, there were generational terms—one set applied to males, and another to females.

Sound confusing? In fact, individuals had their own ancestral lists of names to learn, taught to them by their parents. The !Kung did not know exactly how the list was compiled, but the parents knew exactly what the list should be for each of their children. A long individual memory was not required, however, as many of the

!Kung did not know the names of their great-grandparents. This is an example of cultural inheritance doing its job: keeping things simple yet elegant. Speaking of culture as being simple yet elegant, let's turn the page and see what cultural inheritance was like on another continent—Europe—and at another time—the eleventh century.

CHANGE IS NOT NORMAN

In the classic comedy film *Monty Python and the Holy Grail*, King Arthur and his entourage encounter two muddy peasants in a field, pointlessly moving slop from one pile to another with their bare hands. Arthur describes that he became king because the Lady of the Lake "held aloft Excalibur from the bosom of the water." One peasant, named Dennis, questions whether this legitimates a government, and the conversation continues:

Dennis: Listen. Strange women lying in ponds distributing swords is no basis for a system of government. Supreme executive power derives from a mandate from the masses, not from some farcical aquatic ceremony.

Arthur: Be quiet!

Dennis: Well you can't expect to wield supreme executive power just 'cause some watery tart threw a sword at you!

Arthur: Shut up!

Dennis: I mean, if I went around saying I was an emperor just because some moistened bint had lobbed a scimitar at me, they'd put me away!

Arthur: Shut up! Will you shut up?!

The joke pivots on the unexpected knowledge of the mud-heaping peasant, but why should it be unexpected? What might seem at first glance to have been British "peasant" culture was rich with cumulative knowledge. Historian Peter Ackroyd argued that the Normans, occupying Britain after 1066, co-opted traditional Anglo-Saxon knowledge and landholding systems. Although the Normans introduced French words and built stone castles across Britain—the Tower of London, Chepstow Castle in Wales, Durham Cathedral, and scores of others that still stand—the Norman legal order, according to Ackroyd, was co-opted from the existing kinship-based land tenure system of Anglo-Saxon chiefdoms. Indeed, William the Conqueror made use of this system, ordering his staff to record his new landholdings in detail in the *Domesday Book* of 1086. The king sent personnel across England to find out how much plowland was in each shire, how many livestock and slaves were there, and their value. Hampstead, now one of London's wealthiest boroughs and the home of celebrities such as Ricky Gervais and Helena Bonham Carter, was valued at merely £2.5 in 1086 (one villager, five smallholders, one slave, three plowlands, and a hundred pigs in the woodland). Over thirteen thousand places are mentioned in the *Domesday Book*, mostly rural, and Ackroyd maintained these records were too detailed for William's French-speaking staff to have recorded solely on its own.

The Normans clearly understood that indigenous knowledge was crucial for colonists to survive. Besides, why reinvent the wheel? The Anglos and Saxons had colonized Britain centuries before, introduced Christianity, fought off Viking raids, and inherited some of the ways of the Iron Age tribes before them. At the most elite level, Anglo-Saxon women and men could inherit wealth, and the result was strong, interconnected webs of alliances in both material and spiritual wealth. Here is one example from southern England: in 968, Ælfheah, who was ealderman of Hampshire, donated Batcombe, which he'd received from King Edmund, to his wife, Ælfswyth, a relative of King Edmund, who gave it to their son, Ælfweard, who later gave it to Glastonbury Abbey to care for the souls of them and their ancestors. Ælfheah also gave estates to his brother and nephews, plus the wife of King Edgar and her children as well as Benedictine abbeys.

Follow all that? The point is, Ælfheah was adding value to his kinship ties over two generations. Kinship is one way that culture can hold a tight grip on change. Notice the limited choice of first names in Ælfheah's family. Today, if he reserved a room for Ælfswith, Æfweard, and Ælfheah by phone, the hotel would probably assume there was one person with two nicknames, and a *Python*-esque conversation would ensue. In medieval Europe, complex genealogies traced elite families all the way back to a founding, often-mythical, male ancestor. Like an intricate knot, the Normans, who were small in number compared to the Anglo-Saxon population, had to work within it rather than replace it outright.

It's not just kinship but also material technology that, to individuals of traditional societies, would have seemed timeless. If we go back in time to Neolithic Anatolia (modern Turkey), about eight thousand years ago, we might visit one of Europe's earliest farming villages, Çatalhöyük. Here is where people lived in mud brick buildings, crammed on top of one another. They rebuilt their houses about once per century, literally on top of each other, and replastered the walls up to ten times a year. Family ancestors were buried under house floors, and living spaces were arranged virtually the same from one generation to the next.

A millennium later, Neolithic farmers rolled over thousands of square miles from Hungary all the way to France. Their material culture was so consistent that it is simply called *Linearbandkeramik*, which means pottery with lines on it. In the dense boreal forests, these early farmers built wattle-and-daub longhouses on large wooden pillars, usually with three sections, with trash pits outside the house walls. They cleared fields for the cultivation of wheat, lentils, and peas and built pens for domestic pigs, sheep, and goats, and took the cattle out to summer pastures. In cemeteries, everyone was usually buried so similarly—crouched on the left side, head pointing east—that skeletons actually crouched on the *right* side stand out in contrast.

Inundated as we are with videos, social media updates, and short bursts of novel information, we might well ask, "How would I not go nuts in the Neolithic?" Imagine the husband-wife conversation during the 180th replastering of their house at Çatalhöyük, composing thoughts about lentils and cattle. To a person with a Fitbit and an addiction to *Minecraft*, Neolithic life must seem unimaginably banal, but of course to a person in the Neolithic, celebrity gossip on TMZ.com would probably have seemed equally asinine.

Even for anthropologists trained to appreciate other cultures, the intensity of a kin-based subsistence society can sometimes be too much. In the 1930s, British anthropologist E. E. Evans-Pritchard, encamped among the Nuer pastoralists of Sudan, wrote that they "visited me from early morning till late at night, and hardly a movement of the day passed without men, women, or boys in my tent," to the point at which the "constant badinage and interruption ... imposed a severe strain" on the exhausted anthropologist. Certainly he understood the logic: at the center of all Nuer interests were cattle, the common currency of wealth, social relationships, religion, and daily activities. "Start on whatever subject I would," Evans-Pritchard noted, "we would soon be speaking of cows." The Nuer were "influenced by their love of cattle and their desire to acquire them," held "profound contempt" for people without cattle, and therefore had many questions for him on this subject. Cattle and human ancestors had been intertwined for thousands of years.

Lack of perceptible change was not a problem in traditional life. The change, over thousands of years, from roundhouses to square houses is actually a big deal in Neolithic archaeology. For us, such a time scale can be baffling: thousands of years go by, with seemingly little or no innovation. Nowadays, entertainment and technology are in continual flux, but as Joseph Campbell described in *The Hero with a Thousand Faces*, the core themes remain the same. The classic story of *Little Red Riding Hood*, for example, is at least two thousand years old, and others are even older.

Folktales and agriculture are just two illustrations of the great antiquity of what we tend to think of as "modern." Britain retains many features from the Roman occupation, including place-names (towns ending in "chester"), houses built like Roman villas, and roads directly laid onto Roman roads. Along Hadrian's Wall in northern England in the first century AD, the families of Roman soldiers lived at forts such as Vindolanda, where certain well-preserved remains of daily life could easily pass for modern items. A preserved woman's sandal in the Vindolanda Museum, for instance, with its leather upper and outer sole, nailing, and stitching, looks contemporary, sleek, and stylish. Citizens of Roman Britain played dice and board games similar to modern ones, including backgammon. Traces of writing on a preserved Vindolanda tablet reveal a woman's invitation to "the celebration of my birthday," and another tablet sends for more beer.

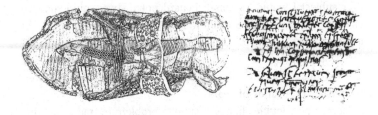

Elsewhere, the antiquity of contemporary culture runs even deeper. At Harappa, Pakistan, forty-five-hundred-year-old terracotta bangles resemble those for sale in the bazaars of the modern-day town as well as the bangles that are popular all over South Asia, some of which are still crafted out of clay. Also, wheel-thrown pottery from ancient Harrapa, dating to around 2300 BC, is just like the pottery that the local Pakistanis use today. That's four thousand years of tradition in jewelry and pottery.

SERIOUS PARTYING

Food and diet are perhaps the most resilient traditions of all, including party supplies such as the beer we drink, which had its origin thousands of years ago in the Near East, and the cheese we serve, a food that may be as much as eight thousand years old. Pots from Neolithic sites retain the residues of the milk or milk-related products they contained, and ancient cheese strainers have been found from sites around 5000 BC in Poland and Germany. In fact, specialized dairy farming was so resilient (nutritionally, dairy products have it all) that certain populations have evolved a tolerance for lactose because of it. Those generations that ate yogurt and cheese built a lifestyle—a cultural niche—that enabled selection for and the spread of a genetic variant (just one base pair) for lactose tolerance in subsequent generations. This advantage enhanced human survival in Neolithic Europe, and now over three-quarters of northern European adults are lactose tolerant.

Besides languages and cheeses, another tradition that dates back to the Neolithic is wealth inheritance, the earliest evidence of which appears to be inherited access to land. It seems the Neolithic era introduced heritable property (land and livestock) into Europe. In tooth enamel from hundreds of Neolithic skeletons from France to Germany to Austria and Hungary, strontium isotopes—which act as geologic signatures of the place where a person grew up— indicate that men buried with distinctive Neolithic stone adzes used the fertile and productive windblown soil favored by early farmers. Those men who had to go farther for their food, away from these valuable soils, were almost always buried without adzes.

Once the seeds of hereditary inequality were sown, there was no looking back. Through the later Bronze and Iron Ages, hereditary inequality and glamor only grew. In a Celtic burial at Vix, in the Burgundy region of France, dating to about 500 BC, a woman was buried with a large wooden chariot, along with an ornate twenty-four-carat-gold neck ornament weighing a pound. Feasting was the theme. Among the burial remains was an enormous bronze wine-mixing bowl—five feet tall and thirteen feet wide, Mediterranean made, with a molded gorgon's head on each handle, supported by a lioness—along with an Etruscan bronze wine jug and imported drinking cups.

This feasting theme—drinking wine and mead—is also found in a roughly contemporary burial at Hochdorf in southwestern Germany. Here, the "Celtic Prince" was adorned with gold bracelets and shoes with intricate molded designs to resemble delicate embroidery. The bronze couch on which he lay—almost ten feet long—was engraved with depictions of wagon trips and sword dances, and supported underneath by small sculptures of coral-inlaid female figurines. The prince's party supplies included eight drinking utensils of aurochs horn or iron, a gold bowl, and a huge Greek cauldron of bronze, with three bronze lions around its rim. At three feet high, it would have held over 130 gallons of mead. We could go on—about the great Scandinavian boat burials, the Vikings, and the like—but you probably get the picture: wealth, alcohol, cheese, and a Super Bowl–party mentality.

FAMILY FEUDS

Why do wine and cheese parties go back so many millennia? One reason might have something to do with the fact that Neolithic feasting was most likely competitive—that is, it was arranged by one patrilineage (a group in which descent is traced back through the father's side) to attract more followers and humiliate rival groups. This is the essence of the famous potlatch feasts on the Northwest Coast of North America, where the more powerful a chief was, the more salmon and roasted pig he could give away at the feast. At some point, prestige became determined not just by an individual's knowledge and achievements but also by inherited wealth and numbers of followers.

Competition can lead to another resilient tradition: blood feuds. If you binge-watch certain miniseries, or are a Hatfield or McCoy, you probably have some feeling for how violence gets inherited from generation to generation. In most cases, there are well-established traditions of loyalty to one's kin group and treachery toward the others. A blood feud means deliberate killings in revenge for a previous killing, under specific rules that include those for compensation as well as a reconciliation ritual that requires leaders on both sides to agree that honor was satisfied.

In Neolithic Europe, as people began to inherit wealth and control over land in their patriarchal family system, cycles of

violence started to appear, probably between competing patrilin-
eages. Several Neolithic massacre sites document some of the targeted attacks on rival villages or clans seven thousand years ago.
At the site of Talheim in southwestern Germany, over thirty people
were executed—many by a blow from a stone ax to the side of the
head while their hands were bound. Few remains were of women,
suggesting that they were captured instead of killed.

Raiding for cattle and women took an even grimmer turn at the
Neolithic mass grave of Schöneck-Kilianstädten, near Frankfurt,
Germany. Discovered in 2006, the grave contained the remains of
at least twenty-six people, including ten young children, who were
buried along with village debris such as broken pottery and animal
bones. Again, there was a dearth of young women among the
remains, indicating that they had been captured and not killed. The
skeletons revealed the fatal injuries from the attack itself, including
embedded arrowheads and holes in the crania from ax blows to the
head, but even worse were the fractures in leg bones (including half
the tibia) that suggested legs were broken as a form of torture. As
Christian Meyer, who led the study of the remains, put it, it was
almost as if the attackers meant to terrorize others and demonstrate
that they could annihilate an entire village.

We might assume these attacks were repeated in reciprocal retali-
ations through the generations. At least this is the pattern that has
been observed in small-scale warfare from Papua New Guinea to
Fiji, in the Scottish Highlands, and among the Yanomami of the
Amazon. To take a case in European history, in Albania during
the Ottoman rule, social organization was based on family lin-
eages linked by marriage alliances or opposed through feuding. By
the early twentieth century, probably one in five Albanian deaths
arose from feuding. For protection, lineage members lived in
extended property-sharing family households, reaching almost a

hundred people and comprising several married brothers and their descendants.

This blood feud system is so resilient that it can lay dormant for decades and then spring back. For much of the twentieth century, feuds in Albania were suppressed by the Communist regime, which banned private property and religious leadership. After the collapse of Communism in the 1990s, however, traditionally powerful families set out to recover their status and land, and hundreds of feuds erupted in northern Albania involving tens of thousands of people. Even as traditional clan leaders and Catholic priests tried to settle the violence, family status once again became dependent on the ability to defend itself and kill others, with higher-status families offering protection to lower-status ones.

In all these examples of inheritance—names, wealth, or conflicts—related events are integrated through *cultural transmission*, the spreading of ideas, concepts, beliefs, and so on, within and between generations. We now know that cultural transmission is every bit as powerful an evolutionary process as *genetic transmission*—the passing of genetic material between parents and their offspring. Through time, the variation that is generated as a result of transmission errors is grist for two other powerful evolutionary processes—selection and drift—that act on variation to create generations that are often quite unlike earlier ones. Still, we know some patterns exist over great spans of time. Why the difference? Why are some beliefs and ideas more resilient to change, whereas others come and go over a few generations? Let's turn the page and find out.

CHECK THE TRANSMISSION

On humid nights in the Middleton Theater lobby, against the sounds of the ice machine, an old video game in the corner, and the muffled voices of whoever was starring in that month's movie, the manager would tell Alex stories about his life. He said that decades before, an ex-girlfriend tried to have him killed. Two black Cadillacs with no license plates followed him in his Corvette to his house one night, then circled the block several times before leaving. A few days after that, newspapers reported a man shot in a Corvette in a mall parking lot in Madison. "They got the wrong man and never bothered me again."

That's what he said, anyway. As the movies changed, the manager told a lot of stories like this, and in each one his safety was in peril. In one story he was holding off a roomful of electricians with a hammer because he hadn't joined their union. In another, he faced an angry sergeant after he screwed up an infantry drill. One day he reported how the pastor of his Methodist church abruptly told him never to show up again. The manager never told Alex why.

These stories were 100 percent unique to the manager, transmitted to Alex and perhaps a few other employees. We wouldn't expect another person to have the same stories. Unlike the manager's stories, traditional folktales often experience a great deal of transmission, with little change, over long periods of time. As we will see in chapter 4, this was the case with *Little Red Riding Hood*, which was told and retold by parents to children over several thousand years. There was some change in the tale, but it was so slow that the variations—a tiger rather than a wolf in the Asian version, for example—can still be identified as versions of the same tale. Similarly, the fairy tale *Snow White* has numerous variations but is still recognizable. For instance, in the Irish version—known as *Lasair Gheug*—it's a little trout as opposed to a mirror that tells the Evil Queen stepmother, much to her anger, that she is not the most beautiful woman ever in Ireland.

The key to whether something counts as a traditional tale versus a story told in a theater lobby resides in transmission, which in the case of folktales must be highly accurate, otherwise errors and/or embellishments would quickly render the story unrecognizable. There are remarkable illustrations of this kind of fidelity among storytellers across the world. In Rajasthan, the region of northwest India where semifeudal rule lasted into the mid-twentieth century, a caste of *bhopa* storytellers has told the same epic poems for centuries. Told and retold by these Rajasthani bards, the stories are incredibly long and long lived. For example, the massive eighth-century BC Mahabharata epic, which details the Kurukshetra War, is a thousand stanzas long and over six times the length of the Bible. Another epic told by a bhopa, when written down by an observer, ran over six hundred pages of text. The story, about a pastoralist who elopes with a personified goddess, and sets off a caste war that kills him and twenty-two of his brothers, later avenged by his son, probably mythologizes an intercaste blood feud from the distant past. It might take a month of eight-hour sessions to recount one of these epics, but they are so accurately retold that the Scottish historian William Dalrymple found that the version he heard late in the twentieth century differed only by a couple turns of phrase from a version recorded three decades earlier by a Cambridge scholar.

Such long-term accuracy of transmission requires long, often-hereditary apprenticeship. As fathers taught their sons to be bhopas, they had them learn ten lines rote per day by age four. This is what cultural evolutionists Rob Boyd and Pete Richerson refer to as *guided variation*—cultural recipes passed down through lines of descent, which effectively act as cultural agents across the generations. These recipes survive only if they are transmittable and capable of being replicated. Teaching is the cultural "bottleneck" of each generation, which is why language itself is—or has been, at least—shaped by its learnability. Although the innate capability of humans to learn a language is extraordinary and unique—a child learns several hundred words by the end of year two, or about what a specially educated chimp can learn in a lifetime—the iterated learning process shapes the language itself to become more learnable and compositional over generations.

A way to do this—relevant also to artificial intelligence, as we'll see later—is to become compositional, or made up of interchangeable parts. Nicaraguan Sign Language (NSL) originated with the first special education programs in Nicaragua in the late 1970s and now is used by nearly a thousand deaf signers. Inventing the language themselves, the first users of NSL—adults who understood Spanish—expressed their signs holistically. To represent a cat that swallowed a bowling ball walking down a street—in a story that study participants were asked to narrate—signers would wiggle

their hands across the body from left to right. Subsequent generations of NSL users, however, expressed their signs sequentially.
The same cat would be expressed in two separate signs: first a
circling motion to indicate the wobbling, followed by a flattened
hand sweeping from left to right to indicate it was walking. By these
later generations, NSL had become compositional, made up of parts
that could be interchanged like words. Cultural transmission had
shaped the sign language.

TRANSMISSION EXPERIMENTS

To explore this evolution through iterated learning, culture evolutionists have experimented with games that are like the children's
game Chinese whispers, also known as Telephone. In a pioneering
experiment, language researchers at the University of Edinburgh
used what are known as *transmission chains* to show how learnability and structure emerges through transmission itself. First, they
asked people to watch moving shapes on a computer screen and
read their randomly assigned "alien" labels—like "kihemiwi" or
"tuge," for example. The participants were later quizzed on a set of
moving shapes for the correct labels. They had seen half the shapes
but not the remainder. Their answers were then shown to the next
participant, who did the same and passed it on to the next participant. In each round, the image-label pairs were divided randomly
into two groups, and the participants trained only on one, with the
other half left unseen. In other words, all the labels were undergoing cultural transmission, just never all through one person. In a
typical round, a participant would copy the image-label pairs, but
with errors and changes. After just several rounds, and without
any intentional design, the alien language evolved to become more
structured and thus easier to learn.

It's easy to do your own variations of these transmission chains. Record a passage of speech into your smartphone, for instance, then let another person listen to it. Then wait a minute and ask the person to record the same passage. Then play this new version for someone else, ask that person to record, and so on. The quick evolution of the passage always takes a unique path, but inevitably the passage becomes shorter and easier to learn.

Here's another example. In a classroom, we hand one player a piece of paper with the following short description, which references the work of Connecticut College's Joseph Schroeder and his students: "Recently, experimenters found by studying forty-seven captive rats that when they were fed Oreos on a daily basis, they became addicted to the sugar content. 'The sugar hit them like cocaine,' said one researcher, who concluded that sugar is a highly addictive substance." We then give each player a blank piece of paper. With about a dozen participants seated in a semicircle, the first one is asked to read the passage, put the paper down on the floor, and then try to rewrite the passage on a sheet of paper without looking at the master copy. The player then passes the rendition to the next player, who does the same thing. After they finish, we observe how the message changed through variation, transmission, and sorting. In one case, the third participant reduced it to "scientists did experiments feeding Oreos to forty-seven rats, and they became addicted to the sugar." As we might expect, the message always becomes shorter within the first few students in the chain.

CULTURAL ATTRACTORS

In transforming through iterated learning, a message also retains certain elements and loses others. The number forty-seven in our classroom exercise, for instance, was almost always retained to the end, as were details such as cocaine, Oreos, and sugar. It is thought that *minimally counterintuitive elements* like these frequently act as *cultural attractors*—elements that get preferentially retained over generations, whereas other elements get left behind.

If human minds are predisposed to certain types of information—Harvard psychologist Steven Pinker argues that the structure of language reveals how human minds were designed by natural selection—some predispositions will be innate. Almost all languages, for example, have words for the colors black, white, and red, and most have words for green and blue. Thinking of the natural world, it's pretty clear that having words to describe plants, blood, sky, and sea should have some survival value. When humans and other primates see red, it elicits a hormonal response related to aggression—think of blood, a baboon's rear end, or blushing skin. In Olympic combat sports such as boxing and judo, the competitor in the red shorts actually has been shown to have a statistical advantage of winning, all else being equal (even controlling for the colors worn by certain countries in the Olympics and so on). Even outside sports, men are perceived as more aggressive when they wear red than when they wear blue or gray.

Researchers have come up with other categories of cultural attractors. There may be emotional bias, as emotional arousal helps people remember experiences. Disgust is also common; the most shared urban legends and news headlines are often sensationally disgusting. But a major category is *survival bias*, the idea being that humans retain important lessons about the environment, potential threats, and reproductive strategies. Over time, survival information would accumulate in folktales. *Little Red Riding Hood* and *Snow White*, for example, contain injunctions such as beware the forest, the unknown, and greed.

Survival bias would obviously be strong in the design of pre-historic technology, which was transmitted across generations as cultural recipes. The recipe for building a canoe is like a story, and one that enhances survival. Ancient teaching of how to build a canoe—from the shape of the bow to the durability of the fiber cords, narrowness of the keel, and quality of the single large tree trunk that was dug out—would obviously have affected success in fishing, warfare, and colonizing other islands. The canoe is also like a story in that certain aspects of the craft could become more prevalent even if they do not affect survival but instead simply are aspects of knowledge that are preferentially passed on from parents to children. At the submerged fishing settlement site of Tybrind Vig, which lies off the coast of Denmark and dates to around 6500 BC, a preserved dugout canoe was found along with intricately shaped and designed wooden paddles, bone fishhooks, and pre-served textiles. If the features are inconsequential to survival—say, the designs on the wooden paddles—we might refer to them as

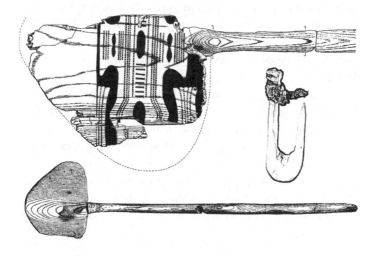

stylistic. They have the potential to change quickly compared to changes in functional features—such as the design of the paddle itself—which affect the mortality or reproductive rates of the people using them.

SOCIAL INFORMATION BIAS

Is there a bias in how social information gets transmitted—not just a bias in *what* gets transmitted but also in *how* it gets transmitted? We would guess so from the social brain hypothesis as well as the human preference for stories with gossip and rumor about family, marriage, sex, friendship, betrayal, social status, interpersonal conflict, and deception. Sometimes we get information from people whom we view as role models. This is often referred to as *prestige bias*—learning from people to whom deference is shown on the basis of their intellect, success, or some other quality or achievement. In traditional societies, these characteristics are frequently possessed by one person, but in a modern world of celebrity endorsements, for example—"I'm not really a doctor, but I play one on TV"—they often aren't, even though social media amplify prestige bias by extending it well beyond the local community, and allowing people to follow international celebrities and public figures through the same action as friends.

Social identity is an obvious attractor. Psychologists have shown that the more polemical the topic—think climate change or gun control—the more likely people on both sides of an issue are to ignore the evidence and go with their preexisting view—their "inner attractor," shall we say. On more neutral subjects, however, people are good at letting the evidence guide their decisions. These polemics remind us that perhaps the most important social information is learning *how* to learn from others, which means learning about cooperation. The benefits of cooperating are obvious at the group level, but they can even be seen for the individual as well, including less violence and shared food, and importantly, also the potential to accumulate cultural knowledge.

How did this cooperation get started in the first place? One theory is *kin selection*, which refers to preferentially helping those people to whom you are most closely related genetically, but this does not explain why people began to cooperate among societies beyond simply their kin groups. Organized religion is a great illustration, where cooperation is predicated not on kinship but rather on oral and written narratives. We may look for both cultural attractors and survival value in comparing religions. Some researchers maintain that small groups that believe in omniscient, moralistic gods who punish are better able to aggregate into larger societies. It has been shown that more complex societies, and hence religions, including Christianity, Islam, and Hinduism, tend to have more punishing and moralistic gods, but this is a correlation without a causal explanation. We'll discuss this in more detail in chapter 8.

Leaving aside cause and effect for the time being, the data show a correlation between religious thought, on the one hand, and generosity and honesty, on the other. In a large cross-cultural study of eight small-scale societies, which included foragers, pastoralists, and horticulturalists, people who believed in a moralistic, knowledgeable, and punishing god shared more readily with each other, including with unrelated strangers who shared the same beliefs. Christians and Muslims exhibited more fairness in economic games than those who held local/traditional religious beliefs, although there were no tests of what effect there was on whom they thought they were playing with or giving to. The more moralistic, knowledgeable, and punishing the shared authority figure was, the more money people typically gave to one another. Those groups whose gods had these qualities were in a better position to exchange between unrelated kin groups, and therefore aggregate and expand at the expense of smaller groups that believed in gods without those three qualities.

In the experiment, the participants were asked to give coins to others according to rules they themselves were able to define. Researchers found that generosity toward strangers was significantly higher—by a factor of five in one test—when both sides believed in a moralistic god that punishes bad behavior and knows your thoughts. They checked this by imbuing local gods with greater knowledge (omniscience of one's thoughts) and/or propensity to punish, but this did not make people more generous. In other words, it had to be a moralistic god that both threatens punishment and knows everything.

What emerged first, generosity or moralistic gods? It's hard to say; meat exchange is as old as our species, but the oldest gift giving or religion depends on how the evidence is interpreted. Was it religion when cave art and lion-headed figurines were made forty thousand years ago, or engraved lines on a mammalian rib fragment a quarter-million years ago? Also, people can be moralistic and punishing regardless of religion; the Nuer pastoralists, as Evans-Pritchard wrote, held "profound contempt" for people without cattle, exchanged cattle to maintain alliances, and rubbed the ashes along the back of a cow to communicate with the spirits of deceased ancestors.

In any case, knowledge and the ability to punish tend to be characteristics of effective leaders. Think of good schoolteachers: if their authority is undermined, or if the teacher is perceived to have the wrong values (like grading randomly), or is judged to lack knowledgeable, the teacher will be less effective. Before *Bad Teacher* with Cameron Diaz changed it, there was a general formula for "teacher movies," such as *Stand by Me* with Edward James Olmos, where the teacher urges everyone to get along, knows not only the subject but also the students, has high moral standards, and can dish it out when appropriate. Looking ahead, we might ask whether artificial intelligence will ever take on these qualities. In many ways, it already has, as an algorithm can be programmed to be moralistic, knowledgeable, and punishing. Take your typical airline booking site, for example, which punishes you for making changes, knows every route and best connection (knowledgeable), and promises the best price for everyone (moralistic).

SLICING AND DICING IN THE DIGITAL AGE

What does all this research on culture transmission say about the future? How might bias, attractors, and morality interact in an era where cultural transmission is of a different magnitude? Let's think in terms of advertising slogans. In fact, maybe there's an idea for a business here, in producing memorable, easily transmittable advertising slogans. The terms "memorable" and "easily transmitted" are important because they lie at the root of what sells and what doesn't. "Just Do It" and "Where's the Beef?" didn't get to be two of the greatest taglines ever by being forgettable. Considering that McDonald's executives are talking seriously about replacing workers with robots, and combining this with the fact that animal muscle tissue can now be grown in a petri dish from stem cells, we might need to get a job to improve one of their slogans, along the lines of "the all-new, robot-fried, stem-cell-grown McDonald's cheeseburger in a petri bun." If they'd hire us, we'd send this message through transmission chains of eight customers to shorten the message and highlight its most attractive aspects through what was retained.

Or maybe we'd simply get our group of participants to start cut and pasting messages electronically. A group of researchers at Facebook considered many cut-and-pasted sentences or passages that appeared on the social media site over a period of years, including this one: "No one should die because they cannot afford health care, and no one should go broke because they get sick. If you agree, post this as your status for the rest of the day." This phrase, exactly, was copied half a million times over a couple years. If this were one of our Chinese whispers–like experiments, we would expect the passage to become shorter. Here is the difference with copying text, though: the next most popular version, which was copied sixty thousand times, was *longer* because people inserted "thinks that" after the sharer's name, and the third most popular version had "we are only as strong as the weakest among us" inserted in the middle. In this social media scenario, cultural evolution reversed its direction by adding words rather than by subtracting them. The opposite can happen as well—the length might be limited to 140 characters or fewer, for example—as social media platforms profoundly affect cultural transmission.

The Facebook team mapped changes in this text between Facebook friends. Not surprisingly, they found that a text's chances of being replicated were roughly doubled when it included a phrase such as "please post this" or "copy and paste," and otherwise boosted by phrases such as "see how many people." But here is another major difference from storytelling: the mutation rate in the text was 11 percent, which means that one in nine users changed it. Now this might be comparable to mutation rates in our Chinese whispers game at the start, but certainly not among our Rajasthani storytellers. Also, that figure of 11 percent was arrived at because the Facebook team identified over a *hundred thousand* different variants of the text among over a million status updates. So while we might talk about a couple of key changes in *Little Red Riding Hood* as it spread into China and Britain over many centuries, adapting to different cultures and environments (such as tiger versus wolf), on Facebook the versions changed, splintered, and conformed to different social groups, like those that turned it into a joke—"no one should be without a beer because they cannot afford one"—and others that transformed into an opposing political view—"no one should die because the government is involved with health care."

One pertinent question comes to mind about text sharing. In our transmission chain experiments, as the participants retrace what happened to the story along the chain, they might share a laugh at a big change because they know how it "should" be corrected. Real cultural transmission is always a group process, with the self-correction feature for group consensus, where learning takes place over a childhood or lifetime. As one bhopa told Dalrymple, "My father used to teach me one story a day, then he would correct me as I recited it back."

How much we can invest in this knowledge transmission depends on the context, though. In the West, where elite college tuition costs the same as a house, children are a long-term investment. This is different from simple agrarian societies, in which children are a net asset in terms of farm labor, and educating them is faster and more easily done by the parents themselves. As cultural complexity grows, the longer parents need to spend educating the next generation. However much you build up in your lifetime needs to be squeezed through the bottleneck of teaching the next generation. We take a closer look at this in chapter 4.

CULTURAL TREES

At the Middleton Theater, the chance placement of objects some-times changed routine practices. For example, when the manager decided to fix his home refrigerator, he brought in all the parts—sheet metal covering, fan, Freon pump, and so on—and stacked them in a precarious pile behind the concessions counter. Under a sign that said "Don't move," the pile stayed there through the summer, half blocking the butter machine, which resulted in lop-sided popcorn buttering. Eventually, Jerry, the regional supervisor from Milwaukee, showed up and ordered the pile of junk to be removed. Had he not stepped in, the pile would eventually have become a fixture. If the Middleton Theater had multiplied as a franchise chain—instead of being razed in the 1990s—a pile of old refrigerator parts might be in some of the spin-off theaters and lopsided buttering in others. Material culture and behavior would have evolved together into new branches.

In the 1980s, a few anthropologists and archaeologists, includ-ing Mike, maintained that stone tools, pottery, and even language are subject to evolutionary processes in the same way that teeth, cells, and bones are. Not coincidently, this was about when Richard Dawkins coined the term *extended phenotype* to refer to inherited traits outside the body. Classic examples include a beaver dam, spider web, bird nest, and termite mound—all of which are "tools" that protect both the organisms and their genes. Those genes, or replicators, are the units that express an organism's behaviors in succeeding generations.

Back then, the academic response ranged from skepticism to ridicule. Pots and arrowheads were tools used by the people who made them, but pots and arrowheads don't breed, so the argument went. Furthermore, prehistoric change in technology, or culture generally, was intentional: humans had ideas, which are in no sense equivalent to genes, and they carried them out. End of story.

Over the next couple decades, however, cultural evolution grew as a field in anthropology and even branched off into many subspecialties that recently reached popular culture. Words such as meme are now widespread, for example, and the idea of technology as extended phenotype is no longer radical. By 2015, over 90 percent of a sample of a thousand Americans aged sixteen to fifty-five considered the Internet to be an extension of their brain, and almost half treated their smartphones as if they were part of their memory. Indeed, as people extend themselves through these devices, online connection becomes effectively obligatory. Since smart devices shape and increasingly constitute our personal environment, they ought to qualify as part of the human phenotype.

THE ACHEULEAN HAND AX

As the popular conversation continues, we occasionally see images of the iPhone juxtaposed with the Acheulean hand ax—the Pleistocene age stone tool that was used by our ancestors from about 1.7 million to perhaps 100,000 years ago. The comparison is a profound comment on the significance of the iPhone, because the hand ax is considered a landmark in human evolution, having enabled *Homo erectus* to disperse out of Africa, and set up shop in Europe and Asia. Some paleoanthropologists argue that the Acheulean hand ax was hardwired into the hominid brain—or that it was at least partly under genetic control. The iPhone has superficial similarities to the hand ax, as both are handheld, multipurpose tools that shape the personal environment. Besides the obvious technological differences, though, there is also a big *evolutionary* difference: unlike smartphones, which seem to change overnight, the hand ax essentially never changed over hundreds of thousands of years. There were slight regional differences, but for all intents and purposes, an Acheulean hand ax was an Acheulean hand ax, no matter where you were in the world.

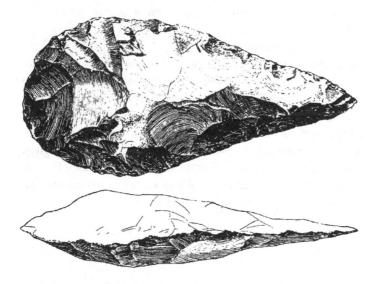

Imagine inheriting a technology that had remained unchanged for hundreds of thousands of years. Our *Homo* ancestors, who probably learned to make stone hand axes as children, would have had little concept of changing this tool. It's all but impossible for us to imagine technological stasis for over hundreds of thousands of years. It's like imagining the distances of interstellar space. How could so little change be possible? The change in Paleolithic tools was not even glacially slow; the glacially carved gorges and waterfalls of upstate New York, for example, are only about twelve thousand years old. Surely, accidental improvements every few generations would have changed the Acheulean hand ax faster than that, yet the archaeological record shows it did not. Why not?

Maybe our ancestors were too stupid to invent anything new. But this explanation is dubious, given that hominid brain size more than doubled during the Pleistocene, from, say, two million to five hundred thousand years ago, and yet stone tools barely changed during that time. If the tools had been constrained by brain power, we would expect changes in parallel with brain size, yet we don't see them.

Alternatively, maybe hominids needed larger groups for technological change to occur. Yet this applies to more complex technologies, where it pays to learn from the expert in the group, or where the group can afford to support technological specialists. In the case of Pleistocene stone tools, probably every individual could knap a hand ax without necessarily learning from an expert—if there even was one. In fact, undergraduate students in archaeology class, given gardening gloves, a pile of flint nodules, and the chance to bang them together, learn the basics of flint knapping quickly, and can soon turn out a passable Acheulean hand ax.

EMULATION VERSUS IMITATION

Given that students can make good progress in an afternoon, perhaps the unchanging hand ax also served as the physical model or blueprint that our ancestors used for making more hand axes. This would be *emulation*, which means copying just the outcome or goal, as opposed to *imitation*, which means copying the method of getting to the goal. The difference is essential to cultural evolution; imitation was critical for complex human technology and culture, whereas primatologists debate whether chimpanzees can truly imitate as opposed to just emulate.

The difference also taps into the mind of ancient *Homo*. We'd like to know when our ancestors started to imitate rather than merely emulate. While the archaeological evidence of the Middle Stone Age doesn't directly show how they learned, it does reveal what they did, which can be reverse engineered from the distinctive patterns of debris they left behind. Find, for instance, a grapefruit-sized flint nodule. Knock off flakes around it to prepare a "core." Rotate the core here, knock off flakes there, rotate, and repeat. Prepare a "platform" on the core. Knock off a flake from that platform. Rotate sixty degrees, knock off another flake, and so on.

To resolve the question of emulation versus imitation, however, we need evidence for how they learned to do it. Archaeologist Jayne Wilkins—who discovered the world's earliest evidence that stone points were used for spears, some 500,000 years ago—reasons that imitators would leave behind similar scatters of stone flakes each time, whereas emulators, with their unique ways, would leave more varied patterns. At Kathu Pan, a site in South Africa dating to half a million years ago, early modern humans made flint blades—long, narrow stone tools, sharp enough to slice up cooked rabbit, use as projectile points, or shave animal hides. Here,

Wilkins believes the variation in debris patterns favors emulation. If Acheulean hand axes were made through emulation, this *could* explain the slowness of change in the Middle Stone Age: the hand axe itself, as the blueprint for its own reproduction by emulation, was under natural selection as part of the extended phenotype of these hominins.

A half-million years later, the first big game hunters arrived in North America, following their ancestors' migration over the Bering land bridge between Siberia and Alaska about 14,000 years ago, when the glacial sea level was about a hundred meters lower than it is now. To make their delicate and refined projectile points, called Clovis points, apprentices would have tried to master an expert's knapping process by careful imitation. Emulation was not an option. Also, unlike the Acheulean hand ax, Clovis points changed in just the several hundred years they were in use, between about 13,300 and 12,500 years ago. Can we order those changes

cm

in time, just by looking at the artifacts themselves? We can, and therein we find insight for mapping technological evolution in general.

EVOLUTIONARY TREES

Evolution is often used as a word for "change," but the true meaning is more specific. Evolution means there are different variants transmitted between generations, over which these variants are *sorted* as some are transmitted more frequently than others. Although first developed for tracking biological evolution, *cladistics*—tracking the history of related entities through their shared features—can be applied to anything undergoing evolution.

This includes technology, so let's look at the phylogeny of those Clovis projectile points. First, we need to choose which technological features to focus on. As a simple example, we can track a single feature, called *fluting*, which is the removal of a long flake from the base of a projectile point. The trees in the figure show three snapshots in the evolution of a projectile point lineage. It begins with the unfluted ancestral state A, which continues its own lineage while also giving rise to ancestor B, the "derived" state, which is fluted. In this new lineage, ancestor B then gives rise to two more groups that are both fluted, which is the "shared derived" state of these groups because they share fluting only with their immediate common ancestor.

With us so far? Let's go one more step. As shown in the third tree, fluting is now old hat for the two newest groups that have emerged, so for these two, fluting is their "shared ancestral state." But if we were talking about these two new groups *and* one of the older fluted groups, then fluting becomes the derived state again because it's what they all share with their common ancestor, B. For reconstructing historical relationships, shared derived traits

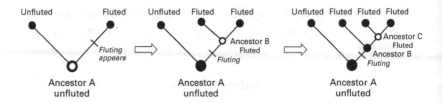

are more useful than shared ancestral traits because they are inherited (derived) from the groups' most recent common ancestor. Now that we have those basics covered, let's see what evolutionary trees tell us about languages, technologies, and maybe even the future.

LANGUAGES AND FOLKTALES

Even more so than technology, human language is strongly inherited and evolves in a treelike pattern over time. Ancestral languages such as Latin branch into descendants such as the Romance languages, which share features derived from their common ancestor, as Spanish, Portuguese, Italian, Romanian, and French speakers would know. Robust language *phylogenies*—treelike representations of inferred evolutionary relationships—exist for major language groups that often reflect remarkably ancient human dispersals. The phylogeny of Austronesian languages reflects the pioneering voyages of ancient seafarers, from Madagascar all the way to Easter Island. The evolutionary tree of Indo-European languages—Hindi, Germanic, and Romance languages and more—parallels many of the human migrations in Europe over the past eight thousand years, from the spread of farming out of the Levant to more recent migrations of Anglo-Saxons and Vikings into Britain.

As these migrating peoples taught their languages to their children, they also told them stories—the same stories they had learned from their own parents. As we discussed in chapter 3, many folk-tales are incredibly old—their time depth being a measure of the faithfulness of transmission. The geographic extent of such stories is tied to the spread of groups across the landscape. In his phylogenetic study of *Little Red Riding Hood*, anthropologist Jamshid Tehrani collected fifty-eight contemporary versions of the folktale from around the world. To break the versions down into discrete features, Tehrani chose plot elements shared among some but not all versions of the tale. In some Asian versions, for example, the villain drinks oil or spring water to clear his throat after he fails to impersonate the child's mother. In certain African versions, the wolf cuts his tongue to smooth out his voice. Identifying the essences of these plot elements, Tehrani gave them labels such as "excuse to escape," "dialogue with the villain," or "hand test" (the children ask the "grandmother" to show her hand through the door). Once he had painstakingly cataloged all these different narrative features worldwide—98 percent of the work, as he will tell you—Tehrani generated a phylogeny and estimated the age of the common ancestor. His estimate that it was at least two thousand years old appeared just as billboards for *Red Riding Hood*, the 2011 Warner Brothers movie, were advertising it as an eight-hundred-year-old tale. Like the pile of refrigerator parts at Middleton Theater, folktales have been around longer than we think and are surprisingly stable.

After that study, Tehrani and his colleague Sara Graça da Silva looked at seventy-six magic-based folktales, including *Rumpelstiltskin* and *Beauty and the Beast*. Their phylogenetic analysis revealed that one tale, about a blacksmith who makes a deal with the devil, was about six thousand years old, making it Proto-Indo European, the ancient ancestor of most European languages and Hindi. Yet rather than blacksmiths, Neolithic farmers, who had no metal tools, were believed to have brought Proto-Indo European to

Europe. Remarkably, a phylogenetic study of ancient folktales has reopened a fascinating debate about ancient technologies.

COMPLEX TECHNOLOGY

Language itself can also be a technology, as with computer programming languages. Tracking the evolution of computer languages since the 1950s, Sergi Valverde and Ricard Solé of the Santa Fe Institute suspected a branching pattern was involved. In the 1980s, for instance, the C++ programming language branched off with object-oriented programming. Another branch emerged in the early 1990s, when James Gosling invented the language that later became Java, which became the world's most popular programming language, especially for web pages.

The figure, which greatly simplifies the analysis, boils the phylogeny down to four major computer languages: Basic, Pascal, Python, and Java. Based on their features, the tree shows that Python and Java are more similar to one another than either is to Basic or Pascal. All four evolved from Fortran, but Pascal, Python, and Java look more like their common ancestor, Algol-60, than they do Fortran. Fortran is what unites those three with Basic. In this tree, Python and Java, together with their common ancestor, C++, form a *clade*. Pascal, Python, and Java, and their common ancestor, Algol-60, form another, more inclusive clade, and so on.

Just as with stone tools or biological species, the phylogeny of programming languages shows abrupt bursts of diversification instigated by key inventions. Zoom in on this phylogenetic tree and you'll find smaller changes at a finer scale, still branching off and nested within one another. If we remove the time scale, the general nested branching pattern could represent another technology, whether it is stone tools, metal weapons, or transistor radios.

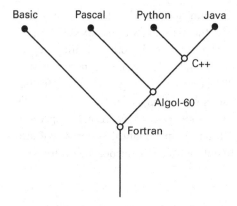

Researchers have explored the US patent database and found the same interdependence of innovations as with the evolution of stone tools, just accelerated a thousand times faster.

A cascade of inventions can often emerge from a novel combination of existing technologies. With a little creativity, this might help anticipate or even shape the future. Given that clades tell us what evolutionary space has been occupied, they might also show what nearby but still-unoccupied concept spaces are ready to be explored. For illustration, let's say Elvis had looked at a cladogram of music in 1953. Phylogenetics wasn't around then, but let's pretend it was. He might have seen a clade for country and western and its closest relatives, and a neighboring clade for rhythm and blues and its closest relatives, with an inviting open space in between. This is not exactly how rock and roll began, but the music did, effectively, colonize open space between existing clades.

In addition to music, we could see phylogenetic trees and tree-like structures being used to examine all kinds of opportunities. For example, César Hidalgo of the MIT Media Lab generates treelike diagrams to represent how closely related technological specializations of different countries are. The closer two products are in terms of branch links, the more likely a country could export the one product given that it exports the other. Unfilled space between two linked products might identify new opportunities.

In the next chapter, we'll take cultural phylogenetics a step further and show how we can use one solid phylogenetic tree, especially a language tree, to infer evolutionary changes in other aspects of culture, such as family structure and political systems. We also briefly look at how this might be used to consider the future.

BAYESIANS

If a certain Saturday night at the Middleton Theater had been his first night working the job, Alex might have thought that the $11 he deposited at the bank was typical. He would have expected even less on Sunday night. After many months of working there, however, he knew that a "new" movie would bring a rush of at least twenty people on opening night. He also knew from experience that the heat and humidity on that Saturday night, given the well-known lack of air-conditioning, discouraged locals from coming in. Also, it was their fourth week showing *Weekend at Bernie's*.

Knowing that business would eventually pick up again, Alex was using what most of us would call common sense, but technically could be labeled a form of *Bayesian inference*. As creatures who develop their own subjective models of the world based on their own unique life experiences, humans use Bayesian inference quite naturally. We continually update our models using new evidence we gather every day. Let's say you own a restaurant and, over time, you've come to know that when you get 30 dinner reservations by four in the afternoon, you'll have about 200 customers for the night, but if you have 70 reservations by six, you'll have about 250 customers by night's end. The longer you've owned the restaurant, the better you probably are at predicting how many customers you'll have.

The general Bayesian approach has several steps. First, quantify your model as a probability distribution, known as a *prior distribution*. Next, gather information about the world you are trying to predict and update your prior. This updated version is technically known as the *posterior distribution*. Now let the posterior become the prior and repeat these last couple steps—collecting information and updating the prior given your new observations—over and over. The refined probability distribution is your result, and it helps you model and predict the world. With respect to your restaurant and the predicted number of customers, you've been Bayesian updating your priors every night, for all those years.

Technically, Bayes's rule—named for the English statistician and clergyman Thomas Bayes, who formulated it—states that the probability of your explanation for what you observed is proportional to two probabilities multiplied together, or namely the probability of the observations themselves and likelihood of those observations given your explanation. It's funny how awkward Bayes's rule sounds when stated formally. Intuitively, we all understand it. Here's another example. Say a tornado hits your house, and you're sure this happened because your sister-in-law put a curse on your family. But hold that thought and become a Bayesian. The tornado itself was a rare event. Besides that, your sister-in-law puts curses on people all the time *without* tornadoes hitting their houses. More formally, we multiply the low frequency of the tornadoes themselves by the low frequency of the cursing being immediately followed by a targeted tornado. Whether you're a human judging these probabilities intuitively or a machine calculating them precisely, Bayes's rule tells you the tornado was probably not caused by your sister's curse.

THE BAYESIAN MIND

What happens when we combine Bayes's rule with the transmission chains we discussed in chapter 3? The answer is, we get not only views of how people think but also a model for artificial intelligence. The "neural conversational model" that Google researchers Oriyol Vinyals and Quoc Le are developing for your next customer service conversation is simple. Instead of handcrafting the conversation rules, like when you book an airline ticket with a computerized agent, their model is uncomplicated and flexible. "We simply feed the predicted output," they write, "as input to predict the next output." The human says "hello!" and the machine responds, "hello!" Convincing so far. An efficient dialogue about log-in details ensues. Later, the human asks, "What is the purpose of existence?" The machine replies, "To find out what happens when we get to the planet Earth." Must be joking. "What is the purpose of dying?" asks the human. "To have a life," replies the machine.

That last response is revealing, if not a little scary. Machine learning looks backward, meaning that its input is the previous output. To feed into its next response, it concatenates what has been said so far. It's called *iterated learning*, and our colleague Stephan Lewandowsky of Bristol University maintains that humans, too, think this way. It's as if you receive a message from yourself, update that message for a moment, and then pass that updated message on to yourself for the next moment. If yesterday the sun rose at 5:45 a.m., and today it rose at 5:46 a.m., you update the model for tomorrow: sunrise at 5:47 a.m.

Lewandowsky's team explored iterated learning with a transmission chain experiment in which they asked people to estimate a phenomenon that hadn't finished yet, like how much longer you'll be on hold during a phone call or what the total gross will be for a movie already in theaters. A participant would read a question such as "If you were assessing an insurance case for a thirty-nine-year-old man, how old would you expect him to be at death?" The participant's response, of course, will be some number larger than thirty-nine. Next, they ask the participant the question again, except the current number was drawn at random between zero and their previous response. So if the life expectancy had been estimated at sixty-seven, the question might be updated to "If you were assessing an insurance case for a fifty-one-year-old man ...," and so on. People answered about one question per second on a computer, with different topics interwoven, until twenty questions had been asked for each topic. Responses from the twenty participants not only converged on the final answer quickly, after about the first five steps, but all the guesses taken together basically matched the distribution of the real-world answers taken together. Through iterated learning, people could make good estimates for all sorts of things—eventual gross sales for a movie, human life spans, the length of poems, the length of reigns of pharaohs, movie run times, and even baking times of cakes in the oven.

Lewandowsky called it the "wisdom of individuals"—like crowd-sourcing your own previous estimates. If we crowdsource lots of these estimates by people making them individually, the "wisdom of crowds" can be an even more precise estimate of the real answer. For the future of cultural evolution, this is important for two reasons. First, social influence—for example, observing someone else's answer—can ruin the wisdom of crowds, and online algorithms are constantly showing us other people's "answers."

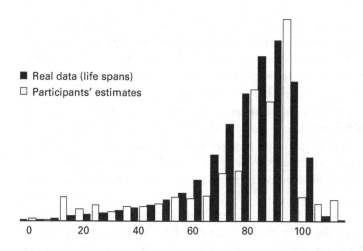

■ Real data (life spans)
□ Participants' estimates

Second, if humans fall short much of the time, it suggests Bayesian inference as a model for an artificial intelligence: update its existing knowledge of a distribution using new information and then sample its new estimate from that updated distribution. We'll see later how this has both promise and problems.

BAYESIAN MODELING AND THE BANTU EXPANSION

Bayesian inference uses the past to predict the future, but it can also be used to interpret the past. If you follow the National Collegiate Athletic Association's Final Four in college basketball, imagine that after the tournament, ESPN asks you to grade the strength of all sixty-four teams, given only the tournament results. All you have for data are the final scores along with the winning and losing teams. How do you grade them all based on that? A round-robin tournament would have let you compare each pair of teams, but the Final Four is one loss and you're out.

You start to panic but then calm down, realizing you can deliver a good result to ESPN. With your laptop computer, you begin by guessing at a strength score for each team. Then you let the computer "play" the tournament, based on the initial setup of teams in the round of sixty-four. Each time two teams play each other in the simulated tournament, the computer picks a winner probabilistically, based like loaded dice on the relative strengths you assigned. You could compare your simulated result to the actual tournament, but even better, you run the tournament a thousand times and compare the most common (likely) result to the actual tournament.

But you've only just started. Your initial guesses for the different strengths of the sixty-four teams are almost certainly not correct. Now you need to guess a new set of strength estimates for all the teams, simulate the tournament another thousand times, then guess another set and resimulate another thousand times again and again, each time comparing the set of a thousand simulations against the actual tournament results. Finally, you accept the set of estimates whose simulations were closest on average to the actual results. This is your answer. Given just one tournament result, you have learned about the relative strengths of all teams and their likely chances against each other. If all that seems like a lot of complicated work, your paycheck from ESPN makes you feel a lot better.

Now if we can use Bayesian inference for a basketball competition, it's not so hard to zoom out and view ancient cultural competition in the same way. To see how this might work, let's go back three thousand years, to western Africa, east of which there was more rainfall and greener savanna grasslands, and even rain forest, in what is now the southern portion of the Sahara. In western Africa, pastoralist speakers of an ancestral version of Bantu began one of the world's epic intergenerational migrations. Over several centuries, the Bantu dispersal ultimately extended southward over a huge swath of sub-Saharan Africa. The dispersal of these cattle herders, who were *patrilineal*, which means they traced descent and rights to property through the male side, culturally bulldozed most horticultural and/or *matrilineal* groups in their path. Left behind was a continent largely populated by Bantu speakers who herded cattle and inherited their lineage identity as well as wealth through the fathers' side. Much of indigenous sub-Saharan African culture today owes at least something to this migration.

Archaeology tells us much of the story of this great Bantu dispersal and the cultures that predated it, including Batwa groups, whose languages are rich with botanical terms suited to forest adaptations; proto-Khoisan-speaking groups, whose descendants include the !Kung San of the Kalahari; and the Hadza- and Sandawe-speaking hunter-gatherers of Tanzania. But let's say we want to learn more from this prehistoric event—something about how cultures change more generally. Is there more information somehow embedded in this record?

It turns out there is, and it involves adding Bayesian methods to the phylogenetic ones we discussed in chapter 4. With respect to Bantu expansion, researchers started with the phylogenetic tree of Bantu languages, which had already been constructed by linguists. They then considered how two specific cultural practices—inheritance and livestock herding—might have changed along the branches of this linguistic history. They divided linguistic groups into four sets: cattle and matriliny, cattle and patriliny, no cattle and patriliny, and no cattle and matriliny.

What next? For one, we need to know the character states at the tips of the language tree for each African linguistic group when it was first described. Fortunately, we can get this information from the handy *Ethnographic Atlas*, which was compiled by George P.

Murdock in the 1960s and 1970s, and recorded essential facts for over a thousand societies. The *Ethnographic Atlas* noted that the Tiv of Nigeria and Cameroon, for example, were patrilineal and herded cattle, whereas the Gangela-speaking Luimbe of Angola or Ndonga-speaking Ambo of Nigeria were matrilineal but also herded cattle.

So far, so good: we have a language tree, and we know which of our four possible character states is at each tip. Now all we need is a model of cultural change that gets us from the original, ancestral state—the "root" of the tree—to correctly predicting the character states at the tips of the tree. What we're after are the probabilities that one event leads to another—that is, if we find the right set of probabilities, then when we simulate the model, it should give us the known result.

Bayesian phylogenetic analysis derives general understanding from just one historical event. The Bantu colonized Africa only once, giving us one language tree and the contemporary cultures at its tips. At each juncture on the tree, a linguistic group will be modeled as existing in one of the above four combinations of character states. Along a phylogenetic branch extending from that juncture, the group can change one character state at a time—let's say from matriliny and cattle to patriliny and cattle. What we want to know is, when one changes, how likely is it that the other will change as a response? Specifically, when Bantu migration introduced cattle herding to a matrilineal group, did this force cattle inheritance over to the patrilineal system?

The answer to the analysis is the set of probabilities of change among the four states. It's a bit like guessing the odds of one basketball team beating another: we guess at the probabilities and then run the model many times for *each* set of them. In the figure, big arrows represent the most likely/common change, and little arrows mean it would rarely happen. Four arrows represent the

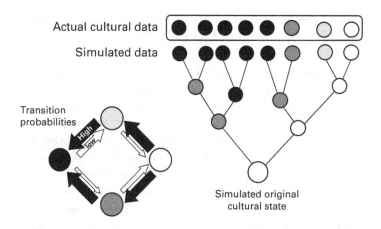

Actual cultural data

Simulated data

Transition probabilities

High
Low

Simulated original cultural state

transitions in one direction, and four more arrows go the other way. Eight arrows represent the eight probability values that need to be guessed.

Got that so far? We have eight arrows representing transitions between four different states. Now start with an ancestral society at the root of the tree—let's say matrilineal and no cattle. At the next juncture, it makes a change: either it gains cattle or switches to patrilineal still without cattle. We choose the change at random, but with probabilities given by our arrows that we assigned before running the model. We roll these loaded dice and go to patrilineal without cattle. The descendants of this new group roll their own dice, this time choosing to gain cattle or go back to matrilineal still without cattle, with the arrows giving us the relative odds. No double jumps are allowed, and staying put is also an option. We keep doing this until we have filled out the whole language tree (remember, we started with the language, which does not change).

Now looking at the end points of the tree, check how well the simulation matches the true states that we know from the *Ethnographic Atlas*. Have the computer run the simulation again and again, maybe a thousand times, just for this particular set of probability arrows. The degree of match with the actual record determines the likelihood that this specific choice of probabilities represents reality. Now change those probabilities, just by a little, and do it all over again. Then again, and again, for different combinations of the eight probabilities (arrows), until we have covered all the sets of probabilities.

In the Bantu study, researchers found the transition probabilities to be quite lopsided. If a matrilineal group acquired cattle, it had a 27 percent chance of becoming patrilineal in the next phylogenetic step. Once a group was patrilineal and herded cattle, it had almost no chance (0.2 percent) of reverting to matriliny. A matrilineal group had a 68 percent chance of losing the cattle if it had them, but only a 16 percent chance of gaining cattle if it didn't.

The cow was "the enemy of matriliny," researchers Clare Holden and Ruth Mace rightly concluded. The wider implication is that introducing a new resource can change or disrupt family life. Working in Ethiopia, for example, Mhairi Gibson and Eshetu Gurmu observed that when piped water was introduced to rural villages—ones where women previously had to spend hours carrying water—two changes occurred. First, fertility (the number of children per mother) increased slightly, and second, younger siblings started leaving their families and migrating to the cities. Although development economists would not have expected these responses to supplying water to villages, Gibson was well aware of the Bantu phylogenetic study, which indicated that family organization is a part of a cultural system that coevolves with resources. To her, there was no difference between adding livestock or potable water to the system.

Another goal is to infer the state of the root of the tree. For instance, what was the kinship system among the ancestors of almost all Europeans? Did married couples live in the husband's or the wife's village? To answer this question, Oxford anthropologist Laura Fortunato started with the well-established Indo-European language tree, the root of which is the Proto-Indo-European language. She then hung the cultural ornaments on the tips of the tree by looking up the kinship system—matrilocal, patrilocal, and neolocal—among the contemporary Indo-European-speaking societies listed in the *Ethnographic Atlas*. After all the simulations were completed, the strongest arrows (probabilities) pointed toward patrilocal for ancient Proto-Indo-European. This fit with independent genetic and archaeological evidence, and it's a remarkable finding.

ACROSS THE PACIFIC

Taking advantage of preexisting phylogenetic trees of Pacific languages, Tom Currie and his colleagues used Bayesian phylogenetic analysis to understand political systems in ancient Polynesia. Starting some fifty-five hundred years ago, the Pacific was colonized by Austronesian people from southern China or Taiwan. This expansion occurred so rapidly that it has been referred to as an "express train." By thirty-two hundred years or so ago, we can recognize in the archaeological record the remains of the so-called Lapita people, who were among the best navigators the world has ever seen. In their famous double-hulled canoes, navigating by the stars and inferring the presence of islands over the horizon by the action of tiny wavelets, Lapita mariners dispersed their culture from Island Melanesia as far as Tonga and Samoa in just a few centuries. In the following millennium, their descendants colonized the rest of Polynesia as far north as Hawaii, as far south as New Zealand, and as far east as Rapa Nui (Easter Island). These colonists brought with

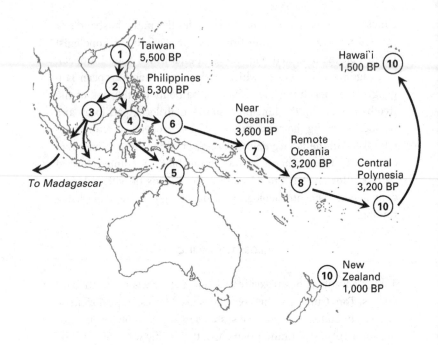

them not only yams, pigs, and chickens but also pottery making, fishing, and their Austronesian language.

Some of the first Lapita colonists of the archipelago of Vanuatu, east of New Guinea, buried a male "leader" with the skulls of three others placed on his chest, possibly sometime after his death or from sacrifice at the time of his death. Isotopic studies show that the man had sailed to Vanuatu, whereas the other three were probably raised there. A few others had the same exotic isotopic signatures as the leader, and most of them were buried with their heads pointing south, which was uncommon in that cemetery, containing over thirty people. Clearly, even this small group of early Pacific colonists already had a set of identities, and probably some social hierarchy, that had to do with where they were from. Hierarchy is probably essential to navigating such enormous distances. When several Mexican fishermen drifted from a mile offshore all the way to Island South East Asia in 2006, they respected their captain during the entire eight weeks, drinking rainwater and eating the raw turtle meat that ultimately killed the American who was on board with them.

The point is, this rudimentary hierarchy among early Polynesian colonists, once left to their own devises on far-flung islands and archipelagoes, ultimately evolved different political systems. On the coral atoll of Tokelau, the prevailing system was *maopoopo*— meaning "together both in body and soul"—with land owned corporately by larger family groups. The Hawaiian Islands, in contrast, comprised highly stratified chiefdoms, with a queen ruling the islands at the time of contact in the eighteenth century.

To figure out how this variety of political systems could evolve from a founder system—the one that buried a guy with three heads on his chest—Currie and his colleagues used our familiar methods. First, they needed a language tree, so they accessed published Austronesian language trees, took a sample of a thousand of the most likely ones, and settled on a single tree that represented the chronological order in which linguistic-cultural groups colonized the Pacific. Next they defined four simplified categories of political systems: acephalous (no head), followed by simple chiefdom (one leadership level), complex chiefdom (two leadership levels), and state. From ethnographic and historical records on the Austronesian-speaking societies of Island South East Asia and the Pacific, Currie and colleagues could categorize eighty-four societies at the tips of the ethnolinguistic tree.

The transitions they simulated—the arrows in the figure—represent how probable a change was in a political system over time. When they compared their resulting phylogeny to the eighty-four Austronesian-speaking societies in recent times, the fit was best if acephalous societies moved up to the simple chiefdom level two-thirds of the time, but fell back again a third of the time. Once a simple chiefdom became a complex one, however, or a complex chiefdom moved up to being a state, there effectively was no going back. The root of the tree was probably acephalous (about 75 percent) but possibly (25 percent) a simple chiefdom. This more or less fits with the chronological position of the guy with the three heads.

Now that we've looked at the ratchet effect of political systems, let's explore the other topic never to discuss at the dinner table: religion. Despite their common origin in the Lapita dispersal thirty-five

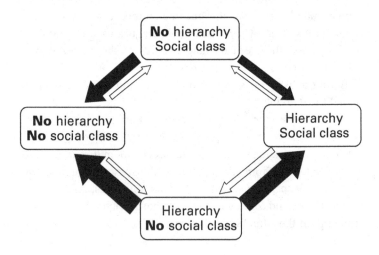

hundred years ago, Pacific societies evolved a remarkable variety of religious practices. One of these was human sacrifice. "The methods of sacrifice," Joseph Watts and his colleagues merrily pointed out in their paper, "included burning, drowning, strangulation, bludgeoning, burial, being crushed under a newly built canoe, being cut to pieces, as well as being rolled off the roof of a house and then decapitated." Ouch.

For their Bayesian phylogenetic analysis, Watts and his colleagues inserted two character states, known from historical sources, at the tree tips: the level of hierarchy of the societies plus whether or not they practiced human sacrifice. They actually had three levels of hierarchy, so they needed to run the model twice—the first time budding low and intermediate levels of hierarchy, and the second time budding the intermediate and high levels of hierarchy. Complementing the study of political change, their analysis showed that human sacrifice "locks in" the jump from low to intermediate hierarchy and then helps it jump up another notch, to high hierarchy. Religion came first, followed by stratified societies. This, too, seems consistent with the ancient Vanuatu man with the accompanying three decapitated heads.

The wider impact of these studies is Bayesian phylogenetic analysis itself. Again, think of it like decorating a Christmas tree: take a historical language tree and "hang" some ornaments of culture on the tips of it. Through simulations on the tree, the goal may be to infer the causal relationship among the ornaments or to figure out more about the ancient root of the tree. Whether the tree is tangled or not—whether its continuities are "vertical" through time or "horizontal" through space—however, makes a big difference in cultural evolution, as we will see in the next chapter.

TRADITIONS AND HORIZONS

One Christmas Day at the Middleton Theater in the late 1980s, Alex set himself on fire while sitting behind the concession counter. *Sea of Love* was showing that day, and Alex's sweater caught fire on the 1950s' electric-coil space heater he brushed up against. He didn't notice the added heat of the flames, but fortunately the manager yelled from the ticket booth, "You're on fire!" Being more or less quick of mind, Alex remembered what he had been taught as a child. He dropped and rolled around on the floor, which was caked with yellow peanut oil, until the fire was extinguished. Once the cheering customers were safely seated, the manager and Alex resumed staring at the same wall, letting the muffled voice of Al Pacino pass through.

Having shown *Sea of Love* for the last few weeks, Alex and his manager knew exactly what to expect from the movie. But this did not prepare them for Alex's sweater catching fire. As we saw in chapter 5, Bayesian inference is great for predicting the predictable. It works in slow-changing or cyclic environments, or when we can create a logical hypothesis about causality. Humans do it instinctually. Jet lag, for instance, is the reestablishing of the circadian rhythm in the brain in a new time zone.

As we've seen, our brains collect shortwave data on behavioral patterns every day. In Bayesian terms, we continually update our "priors," which we codify as cultural norms. This means we won't be surprised that a study of millions of Tweets, as reported in a leading science journal, revealed that people are both happier and tend to wake up later on weekends. As Bayesian thinkers, we already expect this from the workweek, or in Bayesian speak, our priors are hardly updated by the findings. The Twitter analysis, however, gives a future algorithm the same baseline about collective behavior, except at a scale that no individual person could perceive. With this baseline, you can identify anomalies.

Using vast amounts of Twitter and other media content, Nello Cristianini and his group at Bristol University monitored fluctuations in the public mood following specific events, such as economic spending cuts or Brexit. They also monitored seasonal variations in the public mood. They found that especially in the winter months, mental health queries on Wikipedia tend to follow periods of negative advertising. As Cristianini said, people tend to respond to their findings with "That's it? We already knew that!" That's fine with him, as his goal was to quantify human collective behavior rhythms.

Many of these rhythms carry on largely as they have for hundreds or thousands of years. In trying to define what *has* changed, it helps to consider two "shapes" of cultural transmission: vertical and horizontal. The results of these transmission types are what archaeologists call *traditions* and *horizons*. Traditions reach back in a deep and narrow fashion, through many generations of related people, usually residing over relatively small areas. Horizons are shallow and broad and can cover magnitudes more people over a much larger region. Today, because of online media, horizons can easily reach around the world. Sometimes a strand within a tradition can suddenly erupt into a horizon—like in 1976, when Walter Murphy's disco version of Beethoven's Symphony no. 5 in C minor spent one week atop the pop charts. The horizon usually flames out rather quickly, and the tradition reasserts itself. Few people, for example, remember Murphy's "A Fifth of Beethoven." A correction was made, the horizon disappeared, and the tradition carried on.

Let's quantify all this with a rather obscure illustration. In the highlands of Scotland, there is a long-standing tradition of climbing a group of misty hills known as the Munros. There are some 280 of these hills, all over three thousand feet, and for generations people have set out to "compleat" the list by walking up every one of them. These compleatists proudly join the list maintained by the Scottish Mountaineering Council. The cumulative number of compleatists keeps growing steadily—a tradition that consistently adds about two hundred climbers per year. Within this tradition, though, there was a bit of a fad—a horizon—called the Munro Tops, a different group of slippery Scottish hills that tend to be miles from the nearest road, so that a fall might leave you crumpled and alone, writhing in an expanse of soggy peat. During the 1980s and 1990s, the Munro Tops fad rose and fell, as you can see in the figure, while the main tradition of the Munros keeps plugging along steadily, year after year.

With this sketch of traditions and horizons along with how they differ, let's explore in depth a few more examples of longer-lived horizons that have arisen from deep, traditional behaviors around the world, starting with diet and then moving to gender relations and charitable giving.

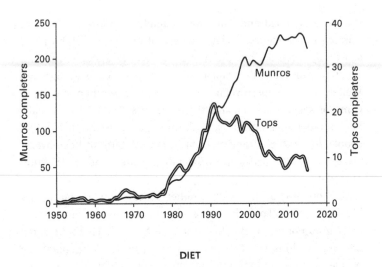

DIET

Often as resilient as languages, dietary traditions tend to move with migrations, because they are learned in families and are part of group identity. About a thousand years ago on the Comoros Islands, for instance, a few hundred miles off the coast of Mozambique, Austronesian colonists from six thousand miles to the northeast not only brought their native diet of mung beans and Asian rice but also then maintained those preferences for centuries across a "culinary frontier" from African mainlanders who cultivated millet and sorghum.

This might sound as if a tradition is becoming a horizon as a result of population movement, but it isn't. Just because people move, sometimes over considerable distances, they can still maintain distinctive food traditions for generations, even when they live near other groups and intermarry. For example, after Hurricane Katrina in 2005, a quarter-million New Orleans residents moved to Houston. Soon after, Andouille sausage, a Louisiana creole staple, began appearing in markets all over the city. Houston saw the same thing earlier with a tremendous influx of Vietnamese refugees, who brought with them their native customs and foods.

Those are some dietary traditions, but into the deep diversity of dietary traditions around the world, one product—sugar—is bulldozing a new horizon. For millennia, cultivating sugarcane was a tradition, after it was first domesticated in New Guinea about eight thousand years ago. Some thirty-five hundred years ago, sugarcane spread with Austronesian seafarers into the Pacific and Indian Oceans. Sugar was first refined in India, and then in China, Persia, and the Mediterranean by the thirteenth century. By the fifteenth century, Portuguese merchants had established large-scale refineries on the island of Madeira. Christopher Columbus married the daughter of a Madeira sugar merchant, and by the seventeenth century, profits from selling refined sugar in Europe drove the slave trade and its plantation economy in the Caribbean and Brazil.

In England, the sugar business took off in Bristol, London, and Liverpool. By the mid-1800s, the leading sugar companies were refining thousands of tons of sugar per *week*. As Bristol engaged with the West Indian sugar trade by the early seventeenth century, slavery was so integral to the trade that after Britain abolished slavery in 1833, the Bristol city council paid £158,000 (about $20 million today) to Bristol slave owners as compensation. Today, the legacy of those profits is quietly visible in stately sandstone mansions—formerly household-scale sugar refineries—and the Tate Britain museum in London, as Henry Tate made his fortune in sugar refining.

What has happened since qualifies as a horizon in dietary change. Westerners now get about 15 percent of their total calories from sugar, which in the form of cane and beets is the world's seventh-ranked crop by cultivated land area. Just three hundred years ago, sugar was not a major caloric component of anyone's diet. Dietary change on this scale has never happened so fast. The last comparable change in human carbohydrate consumption was the transition to dairy products during the Neolithic, but it occurred over millennia. This was long enough for natural selection to spread the gene for lactose tolerance among early dairying populations. In the relatively short time in which refined sugar invaded diets, however, the human body has had no such adaptation. Refined sugar can be as toxic as alcohol. After drinking a can of cola, injecting ten teaspoons of sugar into your system, the pancreas (hopefully) pumps insulin to get the liver to convert your spiking blood glucose into glycogen.

Markets for refined sugar are driving the rapid growth in type 2 diabetes and coronary heart disease, as seen in countries around the world, where each rise in sugar consumption statistically leads to a rise in obesity rates. Americans have doubled their insulin release in twenty-five years. In 1990, about 11 percent of a typical US state population was obese, and no state had higher than 15 percent obesity. By 2014, statewide obesity rates had nearly tripled, and no state was *below* 20 percent in terms of obesity.

Something else happened in 1990: obesity and median household income started to correlate inversely. This meant that poorer households on average showed higher obesity rates than wealthier ones. Starting from no significant correlation in 1990, the correlation grew steadily year by year. By 2015, the correlation was stronger than ever. In states where the annual median household income was below $45,000 per year, such as Alabama, Mississippi, and West Virginia, over 35 percent of the population was obese, whereas obesity was less than 25 percent of state populations where median incomes were above $65,000, such as in Colorado, Massachusetts, and California.

Although this qualifies as a horizon, refined sugar has, ironically, become a tradition among the poor. "I was nine months old the first time Mamaw saw my mother put Pepsi in my bottle," wrote J. D. Vance, who is from Jackson, Kentucky, in his book *Hillbilly Elegy*. Vance, whose poor grandparents moved from Jackson to a steel town in Ohio, describes the traditions of Scotch Irish Americans, underpinned by generations of poverty, from the sharecroppers to the coal miners, to factory workers, to the now unemployed. Besides traditions of group loyalty, family, religion, and nationalism, the Scotch Irish of Appalachia also inherited pessimism and a dislike of outsiders. "We pass that isolation down to our children," explained Vance.

GENDER RELATIONS

Let's move from sugar tolerance to cultural tolerance. It may seem the world has grown more intolerant, with highly politicized social media groups trashing each other and coming up with increasingly irrational conspiracy theories, but as Harvard political scientist Nancy Rosenblum argued, everyday neighbors, in the real world, are still tolerant and cooperative. Political scientists have a theory that whereas the Industrial Revolution disrupted traditional value systems, the postindustrial world is moving toward rational, tolerant, trusting cultural values. As Damian Ruck, a researcher at the Hobby School of Public Policy in Houston, has found, the World Values Survey over the last twenty-five years documents this. Conducted at five-year intervals for over a quarter century, the survey is a remarkable project: lengthy native-language interviews of a thousand people in every accessible country on earth, covering everything from religion to family, civic responsibility and tolerance of others. Among all these values, Ruck discovered that tolerance of others had changed the most, and for the better, over the past twenty-five years.

One dramatic example has been the rapid decline of female genital modification (FGM) in central and eastern Africa. Just a decade ago, nine in ten women in most of central and east Africa—100 million women—had undergone FGM. The tradition has long invoked a cultural perception that natural female physiology is shameful. Women who have undergone FGM as girls frequently must be operated on again to allow childbirth and even permit sexual intercourse. By 2015, the practice had dramatically declined in Kenya: among Kalenjin women, only one in ten under nineteen, compared with nine in ten over forty-five, had undergone FGM. But this horizon is not uniform: FGM is still above 85 percent in Guinea, Egypt, and Eritrea. It declined rapidly in southern Liberia but not northern Liberia.

According to another theory, this global transition—toward self-expression, tolerance of diversity, secularization, and gender equality—is driven by the global decline in marriage and fertility rates. Since 1950, the total fertility rate has declined from 5 percent to about 2.5 percent globally. Marriage is declining even more rapidly. In India, arranged marriage, which has predominated probably since the rise of Hinduism in the fifth century BC, has declined from over half of all Indian marriages in 1970 to only a quarter in 2016. In the United States, despite more ways of getting married and with more government tax breaks, marriage rates have sunk since the mid-twentieth century. By 1960, two-thirds of the "silent generation" was married, but by 1980, it was only half of baby boomers, and by 1997, only a third of Gen Xers. By 2015, only a quarter of millennials was married, and the Pew Research Center even predicted that one-quarter of millennials may never get married.

The fluidity of marriage rates though time and correlation with economics makes it doubtful that humans are "hardwired" for monogamy. Thinking back to the last chapter, perhaps marriage is a cultural adaptation rather than an innate human universal, which has persisted like a large set of branches on a phylogenetic tree with a common root. One theory is that monogamous marriages became widespread during the Neolithic, helping people survive the new threat of sexually transmitted diseases. In fact, the oldest nuclear families identified by archaeology are from the Neolithic. In chapter 2, we described the grisly human skeletal remains of a village raid around 5000 BC. Those remains also revealed a small group whose distinctive isotopes indicated they were from a different village: an elderly female, a man, a woman, and two children. The man was probably the father of the two children, who inherited certain features of his teeth. Was this a nuclear family, with grandma included? Quite possibly; at a nearby site in Germany, two Neolithic parents were buried embracing their children, according to ancient DNA and isotopic analysis of their remains.

If monogamy and nuclear families, never universal among the world's societies, arose as adaptive cultural traditions, they have since needed to be actively maintained. In Europe, where patriliny is thousands of years old, extra-pair paternity (cheating) has

averaged only about 1 to 2 percent per generation during the last four hundred years, according to a study of Y chromosomal data. Similarly, genetic tests in Mali showed extra-pair paternity rates of only 1 to 3 percent among the Dogon. For Dogon women, who viewed menstrual blood as dangerous and polluting, monogamy was ensured by monthly seclusion inside an isolated menstrual hut for several uncomfortable nights. After a woman gave birth, she had to revisit the menstrual hut, and the husband's patrilineage made sure that she slept only with the husband.

The economy of marriages offers explanation for change. Looking through the 2002 census of Uganda, evolutionary anthropologists Thomas Pollet and Daniel Nettle found a supply-and-demand effect: polygynous marriages were more common in districts where women outnumbered men. They also found that polygynously married men owned more land than monogamous ones. In agrarian and pastoralist societies, more land means families need more children as sources of free labor, and children are viewed as wealth.

Fertility is declining, though, in wealthy developed societies, where family economics have been reversed. In a knowledge economy, fewer children mean more wealth to invest in their education. In an age when four years of college costs as much as a house, and many top jobs require further graduate training, the economic choice may be for no children at all. The "DINK" bumper sticker on that Mercedes SL passing you at eighty-five miles per hour sums it up: "Dual Income, No Kids." Compare the household economics of two generations—millennials on short-term contracts, living with parents in dormitory-style complexes, versus baby boomers with long-term jobs and houses that have increased in value twentyfold since the 1970s—and the decline in marriage rates makes sense.

CHARITABLE GIVING

Another trend associated with millennials—however unfairly—is *slacktivism*, a term used by international charities that refers to the facile online support of a trendy cause. Of the million or so Facebook users who "liked" the "Save Darfur" campaign of 2010, for example, over 99 percent made no monetary donation. Some fear that slacktivism threatens the long tradition of sustained charitable giving. They may see a tipping point in 2014, when people posted videos of having a bucket of ice water dumped on their heads in the #IceBucketChallenge (IBC) to support the fight against amyotrophic lateral sclerosis (ALS, or Lou Gehrig's disease). While Charlie Sheen emptied a bucket of money on himself, and a YouTube minister claimed the IBC covertly referenced the Antichrist of the Book of Revelation, regular people donated, often online or from smartphones. Like the Munro Tops we talked about earlier, or the charitable giving after a well-publicized natural disaster, the timeline of IBC donations was a massive burst of giving that waned in a matter of months.

The IBC was immensely successful, ultimately raising over $200 million for the ALS Association—so much money, in fact, that the IBC in effect funded the scientific discovery of a gene that contributes to ALS, with a 2016 paper in *Nature Genetics* thanking the IBC. If the IBC were to become the new paradigm, however, it would pose a challenge: the philanthropist John D. Rockefeller wanted sustained engagement from contributors, who should "become personally concerned" and be counted on for "their watchful interest and cooperation." Rockefeller aimed to establish a charitable tradition. Yet the IBC motivates a horizon model, wherein charities have to continually dream up new, one-off campaigns that quickly spread as a shared trend rather than as a tradition.

Still, it sounds easy enough, so why doesn't it work? IBC spin-off efforts, for example—such as the Mice Bucket Challenge (mouse-shaped toys being dropped onto cats) for animal shelters—were not nearly as successful. In contrast to sustained, traditional giving that is learned across generations, the IBC model might normalize long droughts of funding between large bursts of giving.

Meanwhile, long-running charitable traditions appear as resilient as ever. We can thank millennials, who are no slacktivists; teenage volunteerism in the United States has doubled since 1989. The tens of thousands of family foundations in North America, triple the number in 1980, collectively hold hundreds of billions of dollars in total assets, give tens of billions to charity each year, and are so traditional that they gave only 4 percent less in the year after the financial crash of 2008. The tradition exists outside the foundations as well. By 2015, Americans gave $359 billion—over $1,000 per capita. Even adjusted for inflation, charitable giving has roughly tripled since the late 1960s. The traditional nature of US charity appears also in consistent giving in distinct categories. For at least fifty years, as far back as detailed records extend, religion has been the leading charitable-giving category by a wide margin.

Religion, of course, is the granddaddy of traditions. Despite claims to the contrary, religious traditions abide. Using data from the World Values Survey, Ruck detected only a slow, almost-insignificant decrease in "religiosity" over the past quarter century. Although education may overtake religion as the largest giving category in some future decade, each major category of US charity has grown over the past half century. It is multiplicative growth, where the rich get richer: the more that religious and educational organizations get one year, the more they receive the next year. Unlike horizons, traditions do not go quietly.

It looks as if Rockefeller was right, but there is one catch. Traditions are handed down through people we know or trust, but we might lose trust in these figures, lose our interest in them, or even lose our ability to know who they are. The phenomenon of fake news comes to mind, which we'll discuss later, but another illustration is the university protests around the United States during the academic year 2015–2016. As a result, certain older alumni withheld (considerable) donations to their alma maters, believing that identity politics were getting in the way of solid education. Feeling "dismissed as an old, white bigot," a seventy-seven-year old alumnus of Amherst College stopped his usual donations. In fall 2015, while Mike was dean of arts and science at the University of Missouri, some of the top donors told him they were pulling the plug on their contributions following increasingly nasty student protests, a faculty member slugging a student and not being fired for it (she eventually was), a walkout by the football team a couple days before a nationally televised game, a sudden spike in political correctness, and what alums saw as the coddling of students. On the East Coast, a wealthy Yale alumnus reconsidered his regular (sizable) gift after seeing the viral video of a student yelling at her residential college head, who, with his wife, eventually stepped down from their roles as "house parents." That Yale professor happened to be Nicholas Christakis, a network scientist whose work we discuss in our next chapter. Let's turn the page and take a look.

NETWORKS

One slow evening when the Middleton Theater was showing *Weekend at Bernie's*, the manager told Alex to answer the phone and give a new name for the movie each time a caller asked what was playing. The first call went like this:

"What's showing tonight?"

"We're showing *Weekend at Fred's* at 7:30 and 9:45."

Pause Caller asks again: "What's that title??"

"Weekend at *Ned's*."

The caller's voice got louder, but Alex kept going, under the manager's supervision: "Wait. *Weekend at Jed's*." *Pause* "Wait; I'm sorry. It's *Weekend at Zed's*."

Finally, the caller's voice got very loud, and he slammed the receiver down. With cheeks of beet red, Alex looked at the silent earpiece for a second.

"Nice!" the manager said. "That was hilarious! Let's go back to the normal way now. You never know when Jerry [the regional supervisor] might be calling to test us."

Here we have an example of influence under a simple chain of command: Alex worked for the manager, who worked for Jerry. A dominance hierarchy like this is pretty normal for social primates. "The weak are often exploited by the powerful," writes primatologist Joan Silk. "Strong alliances and lasting bonds are formed; dynasties are established, but are occasionally toppled." Over the past thirty years in Gombe National Park, Tanzania, where Jane Goodall worked, male chimpanzees have been observed to compete over rank and otherwise pursue alliances with high-ranking males. High rank means priority mating access. It also means you can often prevent lower-ranking males from mating at all. Females, on the other hand, have more stable dominance relationships and tend to move up in rank with age, so there is not as much turnover. Daughters frequently rank just below their mothers.

Turnover in dominance is the essence of human drama. "Preferment goes by letter and affection," complains Iago in Shakespeare's *Othello*, "and not by old gradation, where each second stood heir to th' first." Yet dominance networks are just one of many networks of influence at every scale in human society. All people have insiders and peripheral people in their social network—what sociologists call strong and weak ties. We saw in chapter 1 that during their lifetime, a Hadza individual might interact with a thousand or so people but that the number of close trading ties is much smaller, perhaps on the order of six individuals. Those weak ties become essential at a larger scale of society, where gift exchange becomes prone to hierarchy. In antiquity, chiefs competed for supporters through lavish feasts, violence, or precious gifts, which evolved into networks of trade that came to pervade the ancient world.

By 2000 BC, trade vessels crisscrossed the Indian Ocean from East Asia to southern Europe and Africa. Centuries later, the first Polynesians were trading obsidian across thousands of miles of the South Pacific. In the Near East, trade networks—copper from Oman, lapis lazuli from Afghanistan, and blue cotton clothes from Harappa—became foundations for the earliest state societies.

Trade networks manifested into hierarchies of place, from hinterlands to regional markets, to major cities. In the Bronze Age Mediterranean, trade in olive oil, wine, and fish sauce turned otherwise-obscure islands such as Thera into important hubs of maritime trade because of their centrality relative to other Aegean seaports. These hubs were vulnerable, however. If your seaport volcanically erupts sky-high—as happened to Thera in the sixteenth century BC—you're done.

More than a spatial metaphor, networks are everywhere in human life, directing the flow of wealth, information, and ultimately influence. In every modern organization, some network structure directs the evolution of information. Across human cultures, most households have had an inherent hierarchy, reinforced by daily rituals such as eating meals together. A hierarchy usefully funnels expert information from specialists into a community.

Joe Henrich and James Broesch asked people on the lovely Yasawa Islands of the South Pacific to identify their village experts in specific knowledge areas, such as medicinal plants, fishing, or growing yams. Representing each person as a node, they drew an arrow pointing to the expert named by each person. Like a trade network with hubs, each of their network diagrams revealed a hub-and-spoke pattern. The figure shows their network for yam experts, with the size of nodes being proportional to the number of individuals who selected that person as a model. Different shapes represent individuals' villages.

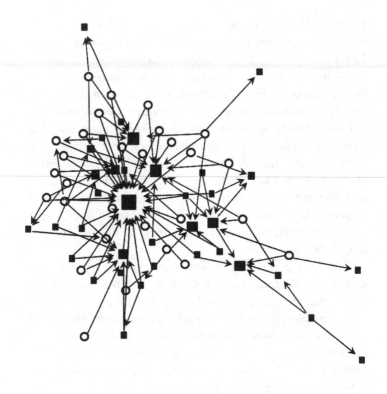

E-NETWORKS

A hub-and-spoke hierarchy is a natural outcome of many forms of network growth, including the early Internet. In the late 1990s, network scientists witnessed the Internet evolve into an elegant hierarchical form, in which most hyperlinks pointed to a minority of websites, and most websites had few links. Without external design, the World Wide Web evolved into something like an airline network, with hubs that allowed any two sites to be linked with just a dozen clicks.

Online social networks also evolved in this way for some platforms but not for others. It depends if you're investing real time or just posing as a website. US teenagers average only 150 Facebook friends each, which, as we saw in chapter 1, is the same as what Robin Dunbar proposed as the stable number of social relationships a person has on average. The distribution in the number of real friends, or Facebook friends, follows a normal distribution, or bell curve, with a distinct average number of friends per person.

Unlike real friendships, Twitter exhibits proportionate advantage, with no characteristic average. The more Twitter followers you have, the more you gain in the future, with almost no constraint. With over 90 million followers, Katy Perry herself follows only 170 people. Like the Internet, Twitter has evolved a highly hierarchical distribution in the number of followers per user. Known as a log-normal distribution, it is the same form of distribution for the number of links per website, the number of followers per Twitter user, or the number of citations per scientist.

Even networks within the Internet evolved hierarchically. Try this game. Look up any subject on Wikipedia and then click on the first hyperlinked word in the Wikipedia entry on that subject. In a relatively short number of clicks, you'll wind up at the topic of philosophy. Starting at the Wikipedia page for "Kermit the Frog," for example, the first hyperlinked word on the page is "Muppets." On the Muppets page, the first hyperlinked word is "ensemble cast," which then leads to "cast members," then "actor," followed by "character," "representation," "semantics," "linguistics," "science," and "knowledge." A few more steps lead to "logic" in the Wikipedia series on "philosophy." As all roads eventually lead to philosophy, we might say it has indirect influence on a vast tree of knowledge.

Online, though, people follow a lot more than Wikipedia. When everyone professes to be an expert at something or other, people struggle to discriminate fake news from real news, science from political punditry, and accomplished careers from professional celebrities. Many blame social media, where everyone has a voice, for flattening what had been a hierarchical information network, paradoxically rendering it simultaneously siloed and globalized.

There is an important distinction to make between social relationship media, such as Facebook and Snapchat, and follower-broadcast platforms such as a blog or Twitter. A Twitter account with just a handful of direct followers can successfully feed information to thousands. Competing against all that content from humans and their chatbots, the magnitude of which we discuss in chapter 9, smart bloggers know how to scaffold their limited direct influence into mainstream media. A prominent political blogger, interviewed in 2016, revealed his strategy for getting his hashtags picked up first by the Drudge Report, followed by a mention on Fox News, and then, with luck, discussion on CNN. In other words, he didn't need millions of followers himself; he just needed what New York University researchers Flaviano Morone and Hernán Makse called "collective influence"—a handful of direct followers who ultimately feed the information to thousands.

Collective influence is not just how many friends are connected but also how many followers of followers, followers of followers of followers, and so on. It captures how true influencers can be all but invisible in terms of node connectivity. Take a major corporation, for instance. Among the half-million e-mails among 156 employees of Enron, the Houston-based energy company bankrupted by internal corruption in 2001, neither Jeff Skilling nor Ken Lay—the two Enron leaders—featured prominently in the message network in terms of node degree or PageRank metric. Having collective influence, without being highly connected, was used in their defense. During the Enron trial of 2006, Lay told the prosecutor that he was part of important decisions only "if I was reachable. Quite often, if I was traveling, they had to go ahead and kind of do whatever they could do." As we know, this "invisibility" defense failed miserably. Both were convicted, with Skilling receiving twenty-four years in federal prison (later reduced to fourteen years). Lay died three months before his sentencing.

A decade later, nefarious online networks elude authority not necessarily by network structure but instead by aggregate behavior. For example, network researchers at the University of Miami used text analysis to examine the European social media site Kontakte, identifying almost two hundred Islamic State of Iraq and Syria aggregates, with over a hundred thousand followers, marked by hashtags such as #khilafah or #fisyria. They showed that each aggregate evaded authority by disappearing for a while and then being reincarnated with a new handle. As the new versions often became bigger than the original, they adapted to efforts to shut them down, continuing the cycle of acquiring followers, disappearing, and then reincarnating.

These aggregates were hub-and-spoke networks, which tend to be resilient against the random loss of followers but vulnerable to targeted attack on their most highly connected nodes. Targeting collective influence, however, works even better. By removing a mere 6 percent of the most influential nodes, Morone and Makse, whom we mentioned above, broke up a Twitter network into isolated, impotent fragments. You'd have to remove twice as many high-degree nodes for the same impact. Their collective influence algorithm was also good at ignoring "fake influencers" on Twitter because high degree is easier to fake then high collective influence. We can see the new potential for online cat-and-mouse coevolution between users and policers.

Since collective influence assumes hierarchy, though, it will be less effective at disrupting highly *embedded* social networks. For a pair of nodes, embeddedness is just their number of mutual friends. For a whole network, embeddedness is the average of that number across all possible pairs of nodes. Unlike hierarchy, embedded networks have multiple alternate routes: if you sever my link with you, I can just reach you through our mutual friend. We keep our friends' friends close, too.

If Enron's Skilling and Lay had been on Facebook, perhaps their special business relationship might have been identified by a simple algorithm. Researchers at Facebook use embeddedness versus another metric, called *dispersion*, to identify special partners in a network. Dispersion is the number of a pair's mutual friends who can be reached only through the pair. A bride and groom, say, could have both high embeddedness and high dispersion if their respective families know each other only through the married couple. In fact, these two metrics alone—embeddedness and dispersion—can identify a social media user's spouse almost two-thirds of the time and also distinguish close family members from friends about three-quarters of the time. Change in embeddedness versus dispersion can even predict the probability that partners will separate.

Whereas hierarchical networks are good at sorting information, nonhierarchical ones, such as small groups, feed into large populations and encourage the random drift of information. Drift is good for fake news and conspiracy theories. Poor at filtering, highly embedded social networks also favor the indiscriminate spread of information through redundancy. People frequently need to hear the message a few times, or at least sense that most of their friends have adopted the idea, before they adopt it themselves. This is what the pop-up message "eighteen people are looking at this hotel right now" wants you to feel.

Researchers have quantified this kind of online conformity among sixty million Facebook users, who were shown two different kinds of banner ads encouraging people to vote. One version showed all their friends who had voted, and the other had the same message but not showing any friends. What is surprising is not that the social "nudge" helped but how little it helped. The social message boosted the rate of clicking "I Voted" by only 2 percent over the information-only message. Across a range of studies, other nudging efforts rarely increase response rates by more than 10 percent, and usually less.

Yet changing the actual social network can transform the spread of ideas. This was shown by an online innovation game played by pairs of players who were connected within a network. In each round of the game, a random pair of connected players was shown an image of an object and asked to provide, independently, a name for it. If the two names matched, the pair won points. In the next round, each played again with new partners from their network. As players saw which names were scoring points in their network, they would often pick those successful names. After multiple rounds of play in a heavily embedded network, where only neighbors were connected, different names became locally popular in the network. With random pairings, however—effectively a nonhierarchical, global network—a single name would ultimately sweep across the entire population, crushing all variation.

One lesson from the random network is that the idea that sweeps through the population is nothing special; any idea can win in the next run of games. Also, the source of the influence—whoever first invented the winning name—could be anyone. If this sounds like online memes, it should; over the years, Microsoft's Duncan Watts has shown this in experiments and models: if people make decisions based primarily on conformity, then a random innovation can occasionally sweep through the social network like an unexpected wildfire.

The critical mass effect also means that passive slacktivists are not useless after all. Slacktivists catalyze the spread of innovation, online or on the ground. They tend to be informed, if not proactive. MIT sociologists showed that in rural India, "passive participants" were actually crucial to microfinancing being adopted by a village, even when the village leader was on board and promoting microfinancing among the village women.

INFLUENCE VERSUS HOMOPHILY

Though eighty years old, network studies have only recently stormed the social sciences. In just a couple generations, the trend in behavioral science has shifted from the classically rational omniscient actors to flawed strategists of behavioral economics, to the social butterflies of network science. In their book *Connected*, Nicholas Christakis and James Fowler argued that everything from happiness to nonstop giggling spreads through social influence. Their most consequential claim has been that obesity spreads this way. As Christakis implied in a TED talk, becoming obese carries the personal responsibility of influencing others, helping to make *them* obese.

But what if it's not influence at all? Christakis and Fowler showed that a person is 57 percent more likely to be obese if a friend is obese, but does this mean they must have influenced one another? What if they live in the same neighborhood, facing the same food choices, all of them fast foods? The clustering of like-minded people is called *homophily*. What if people are merely self-aggregating into "echo chambers" of like-minded people through polarized social media networks and personalized news feeds? In digital markets, advertisers target their messages to niche categories of customers rather than broadcasting the same message to all. In academia,

ResearchGate and Academia.com introduce researchers to like-minded peers, clustering similar academic interests in the network without necessarily involving influence. In business, LinkedIn serves the similar purpose of aggregating as opposed to influencing people.

Usually if we have only a static network, we'll never be able to distinguish true influence from homophily. We need to observe people actually influencing each other, not just clumping together in a social network. One of the best studies was done some years ago by Sinan Aral, who obtained time-stamped data on a network of twenty-seven million social media friends who downloaded the same instant messaging app from Yahoo! called Go. Plotting adoptions in the network through time, plus matching characteristics of users to distinguish similar preferences from social influence, Aral and his colleagues found that about half the adoptions of Go were the result of influence and the other half of homophily.

To really identify influence, however, the gold standard is to observe it in real time. This brings us back to chimpanzees. In the Budongo Forest of Uganda, where the Sonso chimpanzee community has been studied since 1990, chimpanzees regularly used leaves to collect water—enough so that primatologists considered it their universal behavior. Yet on November 14, 2011, an alpha male made a sponge out of moss to get water out of a water hole. As he invented this new technique, he was observed by a dominant female. Over the next six days, seven other chimps from the community made and used moss sponges. Primatologist Catherine Hobaiter caught it all on video. She and her colleagues drew a network diagram with arrows to represent one chimp observing another doing a behavior, along with an indication of when each individual chimpanzee adopted the new behavior. Social influence is visible in the time-elapsed network of each chimp's conversion to moss sponging following direct observation of another chimp doing it.

Real-time analysis was the missing element in mass observation. Soon it will be much easier to identify genuine social influence. Vedran Sekara and his Danish colleagues got downright microscopic by following a hundred first-year university students through Bluetooth on their mobile phones, which can report physical proximity of users within ten meters of each other. These Bluetooth data showed the same people meeting in the same places—classrooms—on weekdays and engaging in recreational exploration on weekends. Nothing revolutionary there, but by updating their colocation data every five minutes, it got a bit better (or creepier, depending on your perspective): identifiable groups emerged from the Bluetooth data revealing people colocating for a time, even if their meetings were ephemeral. Within each group, certain core members attended all or most (75 percent or more) of the meetings, whereas others attended only occasionally (often less than 10 percent of the time).

Now there is not only a network to analyze but also change in that network. That's all Sekara's team needed to predict that a meeting would soon take place if two core members of a group had just gotten together. Also, the closer a meeting approached, the faster the text messages and phone calls were traded among core members, giving researchers a way of predicting when a meeting would occur. Can we predict more than this—say, conflicts rather than meetings? We'll come to this in chapter 10, but first, let's look at prediction in general.

HINDSIGHTED

As far as Alex and his manager knew, a man in Milwaukee owned the Middleton Theater. Given that renovations and even casual upkeep were nonexistent, he apparently used it as a tax write-off. Although they never saw the owner, the regional supervisor, Jerry, often stopped by unannounced. At these surprise visits, the manager would drop whatever he was doing, pull Alex off his stool, and give him a broom, yelling "Jerry's here!!" The manager, a former army private, used to lecture Alex on being prepared, but it was obvious that the surprise visits left the manager with a tangle of nerves. Being prepared was not high on the list around the Middleton Theater.

Speaking of being unprepared, 2016 was a rough year for prediction. You might have noticed. Consider that on June 24, 2016, when 52 percent of the people in the United Kingdom voted to leave the European Union, few, if any, in the British government were ready for it. Rather, they were stunned because almost every poll had showed the United Kingdom remaining safely in the European

Union. One pro-"Leave" member of Parliament reportedly said, "The Leave campaign doesn't have a post-Brexit plan. Number 10 should have had a plan." The member was correct: 10 Downing Street did not have a plan, aside from the prime minister immediately tendering his resignation. Not much of a "plan," but when every preelection poll indicated the Brexit strategy would fail, why do much planning?

Meanwhile, in the United States, the major media were busy leveraging every presidential election poll into highly precise yet ultimately inaccurate predictions for the November presidential election. Most prognostications had Donald Trump with less than a 15 percent chance of winning, and several, including the HuffPost Polster, had him with less than a 5 percent chance. FiveThirtyEight's Nate Silver, who in 2012 scored a perfect fifty out of fifty states in terms of how they would vote, did much better than most, but even on the eve of the 2016 election he had Hillary Clinton with a commanding sixty-five to thirty-five lead. "If you want to put your faith in the numbers," proclaimed the (pro-Clinton) Huffington Post on November 5, three days before the election, "You can relax. She's got this." Uh, got what? After the election, everyone asked how the "science" could be so wrong. How could major polls, including Silver's, underestimate Trump's performance by a percentage point or more in *thirty-plus* states?

Amid all the data-driven probabilities, filmmaker Michael Moore published an essay in July 2016 predicting that "Donald J. Trump is going to win in November." His essay discussed his reasoning, including a neglected working-class population in former industrial parts of the country, elite politicians, and other reasons, which were similar to those that led to Brexit. Moore was basing his prediction on his qualitative theory of *causation*, an anachronism for many big-data analysts. "Causation is for other people," the former chief analytics officer for New York City, Mike Flowers, famously said in 2011. "We have real problems to solve. I can't dick around, frankly, thinking about other things like causation right now." Guess we know where he stands.

PREDICTING THE PAST

How did we get to this point? How will it affect cultural transmission if people base their decisions on aggregate data rather than on personal experience or a theory of causation? Let's go back to 2009, when researchers at Google were already using Google Trends to help predict breakouts of influenza, travel planning, and house prices. In the case of automobile sales, they had a causal model, and it was straightforward: a 1 percent increase in Google searches for "Ford" predicted a 0.5 percent increase in sales of Fords. Later, researchers at Yahoo! showed that searches about certain NASDAQ-100 stocks preceded correlated changes in their trading volumes typically by a day and never by more than three days.

Investors have always talked about the "mood" of the market, so by 2010, when there were already tens of millions of Tweets per day, it was not too surprising to see the paper "Twitter Mood Predicts the Stock Market" appear. Johan Bollen and his colleagues at Indiana University looked at Twitter content from six word bags—"calm," "alert," "sure," "vital," "kind," and "happy"—and found a stretch of time, between December 1 and December 19, 2008, when "calm" words on Twitter had an accuracy of around 87 percent in predicting the daily ups and downs of the Dow Jones Industrial Average (DJIA). Bollen's group used a statistical method called Granger causality analysis, which tells you whether changes in one time series—say, Twitter or the DJIA—consistently precedes a corresponding change in the other time series. If so, we say the first series "Granger" causes the other, meaning that although A precedes B, it doesn't necessarily *cause* B. It might—and in many instances, probably does—but that's not the same as demonstrating cause and effect.

Let's take a look at another study, which found that every three-week change in Google search volume for "debt"—which you can look up on Google Trends—would have predicted changes in the DJIA following those three weeks. If you had used this as your investment strategy between 2004 and 2011, you would have made a 300 percent profit. A plausible causal model was offered for the phenomenon—namely that periods of lower prices are preceded by periods of concern, captured by searches for "debt." Here the strategy is to find the predictive pattern first and then see if the causation seems plausible. Following on from the "debt" study, researchers used Wikipedia to categorize a huge number of words into word bags to do with politics, business, sports, religion, and the like. They found that for the period 2004–2012, the Google search volume of political or business topics—but not the other topics—could have predicted large stock moves five to ten weeks later and made them some money.

Big-data analysts obviously delight in replacing causation with data-driven correlation, but critics jump on such a substitution. A particularly dogged critic of the Bollen word study, using the Internet pseudonym "Lawly Wurm," pointed out that the test period of fifteen trading days might have been cherry-picked to maximize the prediction. Also, since the measure of success was correctly picking a coin flip (DJIA up or down) thirteen days out of fifteen—86.7 percent correct—Lawly Wurm pointed out that the chance of this happening randomly, if tried fifty times, would be about one in six. But how could you try fifty times if you're using historical data, which are real as opposed to simulated? One way would be to test multiple hypotheses: try different word categories and also slide the time window around to capture the most accurate fifteen days in 2008.

PREDICTING THE GAME IN REAL TIME

If after-the-fact explanation is a problem, why not just predict *during* an event? On election night in the United States, there were many prediction machines that you could follow live. As the election results came in, one prominent "prediction meter" shifted in steady, linear fashion from a 96 percent chance that Clinton would win before any results were in slowly down to 4 percent late that night, and finally to 0 percent by the next morning. Back in the day, we called this *watching the game,* not prediction. That said, sports media now cover games with their own dynamic prediction charts. In 2016 college football, Kansas beat Texas (Mike, being a University of Texas graduate, was pretty bummed) for the first time since 1938—this despite having won only four Big 12 games in the previous seven years and following a perfect zero and twelve season the year before. In the first quarter, a prominent sports website predicted Texas's chances to win at 96 percent, and later, with two minutes left in the game and a three-point lead, Texas was still given a 96 percent chance. By overtime, however, when Kansas was getting ready to kick the game-winning field goal, Texas had suddenly plunged to only a 6 percent chance of winning—still kind of a high percentage, given that Kansas was kicking from Texas's eight-yard line. After Kansas kicked the winning field goal, ending the game, Texas was given a 0 percent chance of winning. Seems like a pretty safe "prediction."

By the time a football game is down to a field goal try from the eight-yard line, almost everybody will converge on the same prediction. They are all watching the same game, not playing in it. If they are all playing in the game they are betting on, though, like stockbrokers do, prediction is much harder or even impossible. Think of what Yogi Berra supposedly said: "No one goes there anymore; it's too crowded." In the classic "El Farol problem," modeled by economist Brian Arthur in the 1990s, each person tries to predict how full the El Farol Bar in Santa Fe, New Mexico, will be before deciding whether or not to go. If you think it will be less than 60 percent full, you'll go, but if you think it'll be over 60 percent full, you'll avoid the crowd and stay home. If everyone predicts less than 60 percent, they all go, and if they think everybody is going, then no one shows up. Arthur showed that while the mean attendance converges to the threshold value—in this case, 60 percent—it never settles because everyone is comparing recent results to their predictions of other people's predictions.

Of course, there are lots of ways for human beings to find a good bar, since we are not the mindless agents in the El Farol problem. We have plenty of good algorithms working to solve our coordination problems. They sort our luggage at the airport, protect our credit cards, and help (some of) us with online dating. The algorithms that work on Wall Street are known as high-frequency traders (HFTs). Sometimes HFTs get a bit hyperactive, and lacking creativity or patience, they can get us into trouble. For example, at 2:32 p.m. (EST) on May 6, 2010, according to the Securities Exchange Commission report, automated HFTs were programmed to sell seventy-five thousand e-mini futures contracts, valued at about $4 billion, in twenty minutes. Between 2:32 and 2:45 p.m., other HFTs bought many of the contracts and then traded over twenty-seven thousand of them in fourteen seconds, between 2:45:13 and 2:45:27. With prices dropping by 5 percent in four minutes, the Chicago Mercantile Exchange automatically triggered a five-second halt at 2:45:28, but it was too late to prevent a spread to New York, where the DJIA fell by nearly 1,000 points in under twenty minutes. Procter and

Gamble reportedly fell from $62 per share to $39 before recovering, while Accenture fell from $40 per share to 1¢, and then recovered back to $40. By 3 p.m., the "flash crash" was over, with the Dow recovering to finish down 348 points for the day—still the second-largest one-day drop it had ever experienced.

So much for leaving those pesky HFTs unsupervised! Like with sports and elections, everyone understood what happened after the fact, even if few could immediately agree on exactly *how* it happened. What they *could* agree on, though, was the frightening speed at which events happened, with key events measured in milliseconds: three successive HFT sell-offs reportedly took place at exactly 44:075, 48:250, and 50:475 seconds past 2:42 p.m. that afternoon. Herding occurs in financial trading, but HFTs accelerate it, as if we are fast-forwarding through a zombie movie.

A bit like algorithms, financial traders who send instant messages to each other tend to synchronize their trading activity, as a Northwestern University study showed, and the more traders are in sync with other traders, the more money they tend to make. What we see from all this is that responding to the previous moment, whether to predict the next moment or copy the most recent success, is often the best short-term strategy for the individual. It was for London trader Navinder Singh Sarao, who used spoofing algorithms to make $40 million in the flash crash (he later pleaded guilty to fraud)—but it is not a good long-term strategy for the community or society. If prediction is competitive, then most of us lose, and even the prize for the winners gets smaller and smaller, as predictions become more and more shortsighted.

UNDERSTANDING COLLECTIVE BEHAVIOR

Instead of using big data to predict collective behavior, perhaps we should use it to understand that behavior. Mass Twitter data show the routine rhythms of Internet users: sleeping, waking, swearing, commuting, picking up children and going out at night. Nello Cristianini and his team at Bristol University trained a neural network to recognize different categories of clothing in hundreds of thousands of archived images so that it could distinguish a coat from a jacket, a T-shirt, and so on. How did they do this? By testing a rule on a known pattern, then calculating how far off the rule is, adjusting the rule to get closer, and repeating the cycle about half a million times. When they then applied the neural network to a couple hundred thousand publicly available images of pedestrians walking past a street webcam in Brooklyn, the neural net learned that most people wear T-shirts and short dresses in the summer, sweaters in the fall, and coats and jackets in the winter.

Again, as we pointed out in chapter 6 when we discussed some of Cristianini's finding, one might be tempted to say, "That's it? People wear jackets in the winter? We already knew that." But this is just the beginning. "It is easy to imagine a software infrastructure observing hundreds of webcams," Cristianini and his colleagues concluded in their report, "trying to detect changes, trends, and events." This sounds weird and more than a little ominous—sort of like the television show *Person of Interest*—but before we get too bent out of shape, let's recognize that plenty of people are signing up voluntarily for observation. An example is the "intelligent home." Jason Slosberg, a former medical doctor, is now CEO of LinkBee, which aims to equip homes with light bulbs that include sensors to detect climate, light levels, and pollutants such as pollen in order to improve air quality, energy efficiency, and home security. But it goes much further. By comparing several streams of data from the bulbs, including people's movements, LinkBee wants to infer a resident's physical and mental state. At a central computational center, a neural network tries to work out your mood, or urgent physical condition, such as hypothermia, potential for a stroke, incapacitation, or unconsciousness. When the bulbs detect an anomaly, Slosberg said, "the intelligent home can contact the caregiver to alert them of a potential medical condition."

Even if you don't volunteer to be remotely observed at home through your light bulbs, your health is already monitored in other ways. People tend to Google their symptoms weeks or months before they finally go to the doctor. If people are tweeting a lot of aggressive and stress-related words in a certain zip code, the area tends to have more heart attacks. Using lists of specific words, researchers at the University of Pennsylvania assembled a bag of words related to hostility—mostly nasty, four-letter words—and other word bags for skilled occupations ("conference," "staff," and "council"), interpersonal tension ("hate" and "jealous"), positive experiences ("fabulous," "hope," and "wonderful"), and optimism ("overcome," "strength," and "faith"). Counting up these words by US county, they found that counties tweeting about skilled occupations, positive experiences, or optimism have low rates of atherosclerotic heart disease. Twitter words actually predicted heart disease as well—or even better—than standard risk factors such as smoking, hypertension, or obesity.

So many of these predictions are short term, like our horizons in chapter 6. What about longer-term traditions? The Twitter data tell a traditional story, too, in the interpersonal tension and atherosclerosis across the Rust Belt and into Appalachia. This suggests we might look for a connection between long-term economics and word usage. Maybe the emotion words of a generation, aggregated on a national scale, are correspondingly biased. After all, we have historical eras called the Gay Nineties, Depression, and Fabulous Fifties. Thanks to Google's book-scanning project that started in the 1990s, we have the annual counts of every English word in millions of books over a three-hundred-year period. Counting words in these data from several word bags—"anger," "disgust," "fear," "joy," "sadness," and "surprise," from a resource called WordNet-Affect—Alberto Acerbi and Vasileios Lampos found that emotion words in books declined in relative frequency throughout the twentieth century. This was true for both nonfiction and fiction books. Interestingly, this overall decline was due to a decline in positive emotion words, which had started in the early nineteenth century, with little change in negative emotion words over the two centuries.

This kind of instant feedback among economics, politics, and verbiage is worth discussing. How does it change things? We'll consider that in the next chapter, but we can't help but briefly mention one of the coolest pieces yet of twenty-first-century technology: the Versatile Extra-Sensory Transducer (VEST) technology engineered at Baylor College of Medicine in Houston that translates words via sound sensors into specific patterns of vibration on the body. Its inventor, David Eagleman, predicts that political leaders may someday be outfitted with an Internet-connected VEST when giving a live speech: "Twitter is giving you that feedback immediately. So you're plugged into the consciousness of thousands of people, maybe hundreds of thousands of people all at once hearing your speech, and you can say, 'Ooh, that didn't go over so well.'" Given how much both candidates depended on social media because of the immediate feedback it provides, we would've paid big money to see Clinton and Trump debate each other in the 2016 presidential race while wearing VESTs.

MOORE IS BETTER?

You might recall from chapter 4 the pile of refrigerator parts that lay stacked behind the Middleton Theater's concession stand for months, until one day, Jerry, the regional supervisor, showed up and told Alex and the manager to get rid of them. The manager loaded them up and took them home, leaving Alex to wonder why they were purchased in the first place if they were just going to lie around a theater lobby. Wasn't the goal to actually fix a broken refrigerator? Perhaps the manager inherited both his procrastination and impulsivity—buying expensive parts and piling them up for months certainly counts in our book as "impulsive"—from his parents. About half the variability in behavioral traits such as procrastination, conscientiousness, or impulsivity is genetically inherited. We doubt, however, that the knowledge of how to fix a refrigerator is genetic. Maybe the manager's dad taught him how to fix refrigerators, or maybe the manager was trying to find the time to read a repair manual and figure things out on his own—a form of "individual learning" but with help. Alex never asked him.

On a deserted island, though, even with some prior training, the manager would have had to build a refrigerator from scratch. From Homer's *Odyssey*, to William Golding's *Lord of the Flies*, to the 1960s' sitcom *Gilligan's Island*, Western culture loves to put people on deserted islands and see what happens. Without fail, technology plays a big role in how the inhabitants fare. It has to. Unlike on the original 1960s' television show, shipwrecked passengers on a twenty-first-century *Gilligan's Island* would have swam ashore with more than just accumulated knowledge and a couple-dozen trunks filled with money, stock certificates, and designer clothes. Our modern castaways would have grabbed their smartphones, and it's a good bet that at least some of the devices would have survived the trip ashore. No better way to learn how to build a refrigerator than to look it up on your smartphone—at least until you run out of power.

Think about something else, however: because about a quarter-million active patents relate to smartphones, the shipwrecked group would have brought ashore with them the work of millions of highly technical people, if we assume several people behind each patent. Back on the US mainland, about forty thousand software patents are issued every year. Sorting out all the possible patent infringements, two law professors have calculated, could provide continuous full-time work for two million patent attorneys. This doesn't even begin to account for the millions of people who have populated the Internet with their ideas, views, wisdom, and whatever else is posted there, including hundreds of sites detailing how to build a refrigerator.

The point is, ideas require people to create them—and manage them. Will the island group produce any ideas of its own? On *Gilligan's Island*, the Professor invented everything the group needed—except a reliable boat, of course. He made a lie detector out of coconut shells. He used tree sap to seal raincoats together into a hot-air balloon. He made nitroglycerine from rocks and papaya seeds. But had the Professor died, most of his genius would have gone with him. Future generations of Gilligan's islanders would have plunged deeper and deeper into technological darkness. Fortunately, as documented in a 1978 television movie, they were all rescued before this happened.

THE TASMANIA HYPOTHESIS

Gilligan's Island is a great introduction to a prominent theory put forward by Joe Henrich about how cultural knowledge accumulates over time. Using the island of Tasmania as a case study, Henrich started with the assumption that each person learns from an expert in the population. For a young child, those experts are one's parents, and then later a learner may focus attention on other people in the community who appear to be more successful and/or knowledgeable than the parents about certain tasks. Under this "Tasmania hypothesis," just how skilled or knowledgeable each learner becomes will depend on certain variables and probabilities: how easily an expert can be identified in the first place, how exceptional the expert is in teaching a skill, how accurate the learning process is, and how exceptional the learner is in understanding the knowledge being transmitted.

Under reasonable assumptions, a typical learner usually will not become as good as the expert is, although occasionally a student by chance will surpass the teacher in terms of skill. After Gilligan learned from the Professor how to make use of tree sap, for example, he concocted glue with exceptional adhesive properties that was crucial for building the island telephone. The idea is that with larger populations, there is a better chance that at least one of the many Gilligans will surpass the Professor in terms of mastering some task. At some critical population size, that probability is high enough to expect that in every generation, at least one Gilligan will surpass the Professor, raising the level in every generation. This helps explain why IQ scores have been rising over the last century.

The flip side of the Tasmania hypothesis is that cumulative knowledge will be lost when a population goes through a bottleneck. Prehistoric Tasmania, for instance, fell backward technologically when its population abruptly declined. Technology, though, is not the only thing subject to population bottlenecks. Among Polynesian languages across the Pacific, smaller island populations have experienced higher rates of word loss over the centuries, whereas larger populations have seen higher rates of word *gain*.

The idea that more people means more ideas leads some to take an optimistic view of world population growth, which may reach ten to eleven billion by 2100. Think about it: more people, more ideas. Ideas in the "ether" are a product of our collective brains, linked together through social and technological communication networks. One case study showed that seventy-five years of population growth in a Kenyan town brought about the ingenuity to farm again on what had been barren hillsides, thereby increasing family incomes. As the world's population becomes urban—from an eighth in 1900, to half in 2008, to two-thirds by 2050—there is even more optimism, at least in some quarters.

This optimism is built on two trends. First, population growth is slower in cities, as parents increasingly invest in the education of their children rather than simply having more kids, as has long been a priority in agricultural societies (more kids, more workers). In chapter 6, we mentioned how the total fertility rate has declined globally, but it has declined even further—to below 2 percent—in urban centers. Second, even as population growth has slowed in dense urban settings, the production of ideas, innovation, and information has grown exponentially. Why? Because in large societies, innovation amounts to more than just one idea per person. Instead, it's a dynamic process of idea exchange within networks of people. The pace of urban life grows superlinearly with population size, including the total number of social contacts and volume of communication activity. Both the gross domestic product and number of patented inventions in a city, for example, grow in proportion to its population size raised to an exponent between 1.2 and 1.3. In other words, if population grew from a hundred thousand to five hundred thousand, we'd see the other number jump from three million to twenty-five million.

These are among the "scaling laws" of cities; nevertheless, the communication of ideas still takes place on a human scale. As we know from the television shows *Friends* and *Sex and the City*, people living in urban areas hang out in peer groups. Cities effect little measurable change in embeddedness of social networks—that is, how likely an individual's contacts are also connected with each other. What matters is the *effective cultural population size*, which is the number of people actually sharing information. The productivity of a dense population still boils down to its smaller groups and the fluidity of membership.

For good reason, many managers consider eight people with diverse skills as making an excellent team. A psychology experiment at the University of Arizona showed that a group of eight people, working together on a complex computer game, performed better than an individual or four-person group. A group of sixteen did no better—maybe even slightly worse—than the group of eight. Even per capita, performance was best in the eight-person teams.

Gilligan's Island had seven people with different strengths, including one all-around expert. The Professor oftentimes got help from Gilligan and the Skipper, and sometimes from Ginger, Mary Ann, and (rarely) the Howells.

AN INFORMATION EXPLOSION

In science, the goal is to increase the collective IQ, so to speak, which is why famous simultaneous discoveries are not random coincidences. For example, Charles Darwin and Alfred Wallace, during two weeks in 1858, apparently "discovered" evolution independently. Nowadays, individual geniuses are more difficult to identify because large research teams are the norm. Although it's an outlier, a 2015 physics article set a record with over five thousand authors, and it's not unusual to see articles with fifty or more contributors. As the number of researchers grows—currently at about 5 percent annually—the number of scientific research papers grows as well. The best estimate is about 4 percent per year since 1965, over which time tens of millions of peer-reviewed papers have been published. Those papers collectively cite about a billion references—a number also growing about 5 percent annually. Such growth rates may seem modest, but they are exponential, so the quantities are nearly doubling every decade and a half.

None of this explosion in verbiage is restricted to science, of course, as the sheer number of English words published in books has been growing exponentially for centuries, from millions of words in books in 1700 to trillions per year in the twenty-first century. And these books contain only a few percentage points of the verbiage recorded online. By 2007, humans had already stored two trillion gigabits, with the volume of digital data doubling about every three years. By 2016, more than ten times that much information was stored worldwide—that's the number twenty followed by twenty-one zeros' worth of information. That's a lot.

All this is reminiscent of Alvin Toffler's classic book from 1970, *Future Shock*, which predicted the consequences of "information overload." Forty-five years later, historian Abby Smith Rumsey argued that vast amounts of digital information hinder our collective capacity for forgetting, which is an important behavioral trait that clears away informational clutter, making room for creative thought. As we saw in previous chapters, oral transmission of a story prunes away superfluous details, rendering it more learnable and relevant. In contrast, a viral video gets copied identically millions of times without being streamlined by the transmission process, and actually accumulates more junk in the form of comments and metadata.

Without the kind of vetting that has long typified cultural transmission, culture is bound to accumulate a lot of junk. As technologies crank up the volume, informational junk accumulates on our devices as well as in our texts and videos. A cleaning app can easily wipe a gigabyte of junk data off your smartphone every week—equivalent to a truckful of books or the ancient Library of Alexandria. All junk.

Junk, however, is still part of evolution. Much of human DNA, for instance, appears to be *junk*, meaning that it lacks any observable function. That doesn't mean that it might serve a function; it just means we haven't found one for it. On an evolutionary timescale, natural selection tends to clear away junk about as fast as it's produced by random mutation. Similarly, in early cultural evolution, the size of the community was the constraint on how much junk information might tag along with the more useful and learnable information.

Once we started storing information outside ourselves, though, the constraints were lifted. In this century, there has been virtually no constraint because storage and processing have kept up with the explosion of information. In 2015, the computer giant Intel celebrated a half century of correctness of a 1965 prediction by its employee, Gordon Moore, that computing power per dollar cost would increase exponentially, doubling every two years. This subsequently became known as Moore's law. Just to think how far this could go in theory, César Hidalgo of the MIT Media Lab calculated the ultimate information capacity of Earth—human, biological, and technological—as being about 10^{56} bits (8 bits equal 1 byte). Again, that's a lot. In 2017, the largest computer in the world—in China—held about 10^{15} bits, which, compared to the "planetary hard drive," is like the diameter of a single electron compared to that of a large galaxy. Hidalgo calculated the planetary hard drive has used up 10^{44} bits out of its total capacity, which is like having traveled half a centimeter on a drive from Boston to Seattle.

THE EXPLOSION HITS SCIENCE

Storage has increased in science too, not just in terms of computer data but also in terms of space in peer-reviewed journals. After the online, open-access (meaning that its content is free) journal *PLOS ONE* began publishing in 2006, it doubled in size every year for six years in a row in terms of the number of papers that appeared. As more reviewers have joined the effort, *PLOS ONE* can maintain a rigorous review process even as it publishes about a hundred papers per day. This high-volume model has spread in reputable ways—some prestigious print journals such as *Science* and *Nature* have online journals as well—but it has also given rise to hundreds of predatory open-access journals that will publish, for a hefty fee, almost any academic study on almost any subject.

Like the headwaters of a river branching into the uplands, new niches of science spring up in increasingly specialist soil. As a headline in the satirical newspaper the *Onion* read, "Scientists Make Discovery about World's Silt Deposits but Understand If You Aren't Interested in That." In 2016, there were twenty peer-reviewed papers published on the niche subject of the neutron activation analysis of archaeological pottery. If you are a specialist with an interest in that topic, that's manageable, but that manageability does not extend to the eight hundred or so papers published that year on neutron activation analysis generally, much less the several thousand papers on archaeology. If you worked on something more mainstream, such as obesity—as we do—you have over twenty thousand papers to deal with from 2016 alone. This is beyond human capacity to absorb, which means science may be reinventing the wheel more often: more stuff, but less change.

Let's take a closer look at the diminishing returns effect between the bulk volume of what gets published and amount of new information created. There is a general law of diminishing returns in how the size of vocabulary—the number of different words—increases only with the square root of the raw number of words published: if you increase the number of books by a hundredfold, you increase the vocabulary contained in those books only tenfold. This effect extends to citations one makes to the work of others. Nowadays, most citations are to works that are only a few years old. As the volume of scientific papers grows exponentially, this window of several years represents a smaller and smaller fraction of the total scientific literature being produced. If the volume of papers doubles every decade—a 7 percent annual growth rate—a bibliography today covers only half as much of the expanding field as the same-size bibliography a decade before.

But here's the interesting thing: because of the rapidly expanding number of articles and books being produced, the odds of having your *work* cited increases, especially if it's only a few years old. It used to be, as Derek de la Sola first noticed in 1965, that most scientific papers were never cited—science's coldhearted way of shedding junk. In 1980, about 30 percent of all published research went uncited, but by 2015, only about 10 percent of papers went uncited—a fraction that is still declining. Inflation of one's citation numbers is such that hundreds, if not thousands, of living scientists now have a higher h-index than Darwin does (the h-index measures how many publications a researcher has with at least that number of citations—for example, an h-index of sixteen means that sixteen different papers have at least sixteen citations). And of course, the best way to generate higher citation counts is to cite your own work as often as possible, even if it has nothing to do with what you're currently writing about.

Borrowing from Lewis Carroll's *Through the Looking Glass*, this is the Red Queen effect, meaning that researchers need to do more and more just to stay in one place. Increasing competition to get into the top journals is ferocious, and it's not surprising that there have been some high-profile retractions in the past decade for papers that faked data. One, on the attractiveness of Jamaican dancing, made the cover of *Nature* in 2005. After the senior author later found out that the lead author had faked the data, it took him a long struggle to have the paper retracted, which did not happen until December 2013 with a published two-sentence retraction letter in *Nature*. It seems the publisher was more interested in heavy citation counts than in whether the citations were to admittedly bogus science. In 2015, a *Science* paper faked data on how door-to-door canvassers can influence people's opinions about personal stereotypes. Interestingly, the hypothesis was validated by a genuine study by different scientists the following year.

Dutch psychology professor Diederik Stapel, who was suspended from Tilburg University in 2011, started out by manipulating experimental psychology results and in later years simply faked his data to achieve the kind of "results" that prominent journals love. As a senior scientist, Stapel was at the center of his network of junior coauthors, who were apparently unaware of the fraudulent data. Citations of Stapel's work have naturally plummeted, but his closest coauthors seem to have gone down with him. Guilty or not, reputation matters in science.

WITHOUT SELECTION

With all the junk, inflation, and fraud, can we still find and build on the best ideas, like on *Gilligan's Island*? Without selection, evolutionary drift fills the void: a lot of activity, signifying little. Random drift is variation and transmission, but with little or no selection. The rash of fake news online may come to mind, whereas another well-studied example is first names, formerly traditional as we described in chapter 1, but now subject to random copying and thus evolutionary drift. With random drift in an expanding volume, even though more new ideas are generated, change in the most common ideas can paradoxically grind to a halt. This is because invented ideas, which start from obscurity, are less likely in an exponentially growing corpus to drift up into the popular conversation. Expanding populations have more ideas, but those ideas then have more trouble making it into common vocabulary because the most popular ideas are surfing the wave to ever-increasing popularity.

It's the Red Queen effect on steroids. Paradoxically, exponential growth can make it look like the winners under drift were selected. Think of a "top two hundred" list of the most popular ideas or topics in some genre. It could be Twitter celebrities, trending scientific topics, or just the most popular English words. There is turnover on that list, with new entries replacing others on the top two hundred. Under random drift, there is usually continual turnover on this list, as new entries become highly popular just by the luck of random drift. When the corpus is growing exponentially, however, the rich-get-richer effect at the top keeps raising the bar. In 1970, the top 1 percent of five-year-old research papers had only fifty citations each, but by 2005, the top 1 percent of five-year-old papers had over a hundred citations. As the top 1 percent starts to age, turnover at the top slows down. The same thing is found among English words. In books from the early 1800s, there was a turnover of about seven or eight words per year among the top two hundred words. By 1900, the turnover was only two or three words, and by 2000, the turnover was only one word per year among the top two hundred. More volume, less variety, everywhere you look.

These exponential trends won't continue forever, however. Although the ultimate planetary hard drive may be huge, Moore's law for computers has finally begun to level off, at least for the time being, as transistor size would need to get smaller than ten nanometers for the doubling law to continue. Moore's law is finally entering into a third phase of the classic S curve cycle of invention, innovation, and leveling off. As capacity levels off, will stronger selection return, surfacing all the good ideas again and shedding junk? Will more people mean more *good* ideas again? Maybe, but given the vastness of the informational universe, humans will need help. That help—the new S curve—is artificial intelligence. Selection will be possible again when machines are the ones vetting the information, but it won't be the same, as we'll see in the next chapter.

FREE WILLY

On the days when Alex worked alone at the Middleton Theater, he sold customers their 99¢ tickets outside and their Mountain Dews inside, and then scrambled up to the projection booth to start the movie. Up in the booth, the film ran from one enormous rotating platter, through rollers and across the projection lens, to an identical receiving platter that wound it inside out for the next run. One day, after Alex had started *Free Willy* and was back in the lobby dusting the Good & Plenty boxes, a customer showed up to say the screen had been blank for ten minutes. Rushing upstairs, Alex found the platter pouring film onto a hopeless tangle on the floor. In a panic, he cut the entire tangle out of the film and respliced it, and for the rest of its run *Free Willy* began ten minutes into the movie. Although the customers didn't seem to care, they surely noticed the abrupt jump from the middle of a local advertisement for a "Murcree" car dealership to a scene of Willy the orca swimming around.

The change was sudden and illustrates two points about the future of cultural evolution. First, a gap in an expected narrative upsets our working memory, which is an essential aspect of cultural capacity, and becomes an issue as algorithms and artificial intelligence take on more prominent roles in cultural transmission. Chimpanzees have enough working memory to make complex tools and maintain rudimentary behavioral traditions, such as remembering how to use a tool they haven't seen in years, but humans can use memory for much, much more, such as remembering subprocedures embedded within larger sequences that are themselves embedded in cultural memory. People can carry out complex activities on timescales ranging from a few seconds, such as solving simple algebraic problems, to many years, such as raising children.

Second, feeling disjointed, as with moviegoers watching an ad for a Mercury car dealer in one moment and then an orca swimming around in the next, has become our way of life. As Thomas Friedman put it in *Thank You for Being Late*, technology may already be changing faster than human behaviors, laws, institutions, and customs can adapt. This isn't generational change, as Alvin Toffler described in *Future Shock*, but rather intragenerational change. It's happening in all three elements of cultural evolution—variation, transmission, and sorting—through media that are both diverse and rapidly changing. A 2014 survey of US teenagers, for example, still ranked Facebook as the national favorite social medium, but it was already challenged by Instagram, Snapchat, Vine, Tumblr, and newer platforms. This doesn't count all the messaging apps such as WhatsApp, Viber, and many more that collectively have more users than those big social media networks. Rising and falling in popularity, each new social media or messaging platform imposes its own unique biases on variation, transmission, and sorting, all leaning toward a company's goals as much as the user's.

If we take a step back, this unsettling flux represents a transition between modes of cultural evolution. We'd like to use an analogy that is helpful, if not taken too literally. Let's think of the memory component of transmission as the depth of water in an ocean—shallow in places and deep in others. Our ocean has three different kinds of animals inhabiting it, each representing a mode of cultural evolution past, present, and future. Bluefin tuna are in the deep end, massive schools of herring are in the shallow end, and swimming throughout the ocean we have orcas. Let's take a closer look at our ocean dwellers.

BLUEFINS AND HERRING

Bluefins represent local traditions stretching back deep in time. They can dive down to a kilometer or so, they travel in relatively small schools, and they can collectively remember things such as distant migratory locations. Also, like traditional cultures, bluefins are in danger of disappearing. Herring, on the other hand, move in massive schools—as large as millions of fish—and spend time in shallower, coastal waters. Similarly, within a shallow time depth, algorithms guide human followers like schools of herring, using popularity as a beacon. Tank experiments show how a robotic fish moving unwaveringly in one direction can lead a school of real fish with it, just as we saw in chapter 1 with Ian Couzin's birds. Primates can be led the same way: a few baboons can lead the whole troop to a new food patch if those few travel in the same direction. Animal scientists call it *directional agreement*. Marketers want the same result and employ data-mining companies to amass thousands of data points per person, aiming for feedback among targeted advertising, human response, and more finely targeted advertising to lead consumers in a certain direction.

Clearly, the more that algorithms facilitate human communication and decisions, the more profoundly they change both the tempo and mode of cultural evolution. For online choices, a popularity bias is driven by search algorithms, which effectively rank options by popularity or network centrality. In social media, positive ratings are a prime currency. Popularity increasingly outranks quality. Whether it's a hotel room, an investment, or even a scientific theory, people are more likely to endorse something once others have endorsed it. Sure, quality is part of a rating, but if people are copying each other's errors—which can be on a magnitude scale, where we think a quantity is in the hundreds when it is really in the thousands—the errors do not cancel out. Rather, they feed back multiplicatively into the crowdsourcing algorithms. In 2013, for instance, Google Flu overestimated its influenza estimate—which assumes people Google flu-related terms from their direct experience—because many people were Googling what other people were Googling, which in turn led Google to suggest those search terms, and so on.

In cultural evolution, a shallow time depth means freedom from the deeper past, which often allows more turnover and drift. In 1960, David was the most popular name for boys in virtually all states west of the Mississippi River, with Michael, James, Robert, and John rounding out the top five in almost every state. Now, baby names in the United States are freely chosen—no longer traditional or inherited—and the invention rate has tripled in the last several decades. Turnover in the top hundred is rapid. Sublists guide parents to just the right name for their social group or region—like Addison and Beulah among the best southern names. This has balkanized the landscape of naming. By 2015, neither of the top two boys' names in Wisconsin—Oliver and Owen—was among the top thirty in California. This is typical of ecological drift. Geographic dialects, for example, evolve among the songs of birds, which copy each other with changes arising through recombination, invention, or errors.

Likewise, drift in social media content creates polarized groups. The silos of fake news on social media suggest linked, drifting ideas that bundle into identities. A "vast satellite system" of fake news sites now surrounds mainstream media sites, as Jonathan Albright of Elon University described it. Algorithms help this happen because the network of fake news sites is highly entangled, such that each site already is well connected, which helps it climb up Google's PageRank algorithm, thereby prioritizing sites by network connectivity and popularity rather than by validity.

Similarly, social media and crowdsourcing may be making schools of scientific thought more herring-like. Swimming in shallower waters, in terms of more recent scientific bibliographies, scientists are increasingly crowdsourcing their attention through social media feeds. Algorithms used by Google Scholar, Mendeley, ResearchGate, and Scizzle feed articles to scientists through a personalized balance between the scientist's topical interests and their social as well as citation networks. Other scientists code their own Twitterbots to automatically scan for articles with specialized keywords—in the process attracting hundreds of scientists as followers.

To counterbalance this, the journal *Nature* advised scientists to "go to seminars and meetings," and it quoted a young scientist who opined, "Weekly events can help bring people out of their offices, [and] create a sense of community." If it seems remarkable that anyone would need to be reminded of this, that's because science has already moved decisively toward virtual collaboration among scientists, their algorithms, and the entire scientific record.

ORCAS

Orcas are perfect models for future knowledge grabbers: intelligent, selective hunters that, working individually or collaboratively, choose their prey from anywhere and at any depth down to their absolute limit. Let's see how these qualities might apply to scientists. Ideally, open, online collaboration would move scientists toward building on the latest and most relevant science and away from schooling, like herrings, around citation statistics and network links. An encouraging model is GitHub, the Wikipedia of software design, where thousands of developers openly collaborate on projects. Fueled mainly by validating comments from their peers ("Good job!"), the GitHub community dives deeply and even does free projects for giants such as Microsoft and Hewlett-Packard. On GitHub, "ultimately you have an expert—the person who wrote the original program," noted Thomas Friedman, "who gets to decide what to accept and what to reject." With the best ideas vetted by experts, GitHub exemplifies the Tasmania model we looked at in chapter 9, in which progress is accelerated by a large population size. The massive global population of orca-like GitHub developers can complete projects in a fraction of the time that it might take a team of paid herring-like employees.

As our allegorical orcas dive deep for prey, they encounter millions of scientific articles lying in their cold underwater tomb. As we saw in chapter 9, many of those articles, especially the older ones, have never even been cited. Virtually any article, no matter how obscure or old, can turn up in an orca's search. "Nothing in the past is lost. ... [E]verything exists on one plane," wrote poet and journalist Dan Chiasson, so if you are a researcher with brilliant, but uncited articles, take heart, because it's a good bet that they will one day be brought to the surface. Drawing on digital information both globally and historically, knowledge evolution will be aided by artificial intelligence to select for well-specified qualities. If open science adopts this kind of expert selection, science will shift up a gear. Following the release of the gene-editing technology CRISPR in 2014, Kevin Esvelt of MIT saw open science as morally imperative, especially as humans begin engineering the evolution of animals, insects, plants, microorganisms, and possibly even themselves.

A fully open science can also study itself in order to optimize its own evolution. Forecasting future citations of new medical papers, for example, can help predict whether a drug approved by the Food and Drug Administration might appear a decade later. Deeper insights will come through meta-analyses of text-mined scientific publications. Biomedical researchers at the University of Manchester, for instance, are looking for disciplinary trends and mapping the flow of information between them. Identifying such networks can highlight key new areas of research—potentially those that an algorithm could use to generate a new hypothesis. "Algorithms will build on algorithms," promised a trade magazine at Hewlett-Packard, "with every prediction smarter than the last." Predictive algorithms learn through a process called *supervised learning*, where thousands of successive estimates are checked against the correct answer, each time adjusting the model parameters after each trial to improve the estimate incrementally. This is the essence of Bayesian modeling.

The algorithmic approach needn't be restricted to published literature; it can also study real people. It could use a platform such as Amazon's Mechanical Turk, which now hosts over twenty thousand online participants per month on experiments ranging from rating facial attractiveness to studies of generosity and religiosity. Online social research already has global reach, as over half the world's population has mobile phones. Machine learning can infer much from basic phone data, even the personal wealth of someone in a developing country. Researchers recently compared billions of interactions on Rwanda's largest mobile phone network to personal phone surveys that provided direct estimates of personal wealth. Machine learning estimated the personal wealth from the phone contacts, volume, and timing of calls or texts and geolocations. It could even predict whether a person owned a motorcycle or had electricity in the house.

Anonymous phone data could also be used to predict conflicts. No conversation content is needed. Instead, all you need is the timing of events, as they tend to accelerate in a predictable pattern, like a ball dropped on the floor: bop, ... bop, ... bop, bop, bopitybopi-tybopbopbop. As a conflict escalates, the shortening time intervals between responding events—whether years, days, or seconds—are inversely proportional to the numerical order of the event, via a negative exponent called the *escalation parameter*. The method has been developed on data sets on the ground and online, regarding escalations of warfare as well as online discussions preceding an attack or civil unrest. It even applies to a fight at the family dinner table (you can verify this with a stopwatch and Will Ferrell's classic "I Drive a Dodge Stratus!" skit on YouTube).

After wealth and conflict, the next step is to predict health. The future promises many marvels. Hewlett-Packard says that by 2030, your embedded microchips will alert you when it's time to 3-D-print yourself a new kidney. This is not so farfetched. Even now, a Google user's search activity—for certain diagnostic symptoms of, say, the onset of diabetes or a chronic condition—can reveal a developing health problem before it is even known to the user. Governments are interested too, of course, at the public scale. In the United Kingdom in 2015, the National Health Service agreed to share millions of personal health records with DeepMind, the Google-owned company that developed a neural network called a "differentiable neural computer" that learns to understand narratives, analyze networks, and solve complex logistical problems.

Speaking of which, did we mention that the orca has a huge brain? Neural networks aim to solve problems the same way a human brain would, with layers of networks that discover patterns of patterns. An image of a face might enter the input data layer, which is passed through layers of intermediate representations, like the edges that make a shape, and then the shapes that make a face, to the response layer. A dynamic neural network learns by rewiring its neurons to reinforce whichever millions of neurons that activated—"voted"—for the correct answer. Given that the number of connections in a neural network grows with the square of the number of nodes, the pattern recognition of neural networks can become much more granular real fast.

Nevertheless, we have not yet built machines that truly reason like people. Researchers at MIT's Center for Brains, Minds, and Machines say artificial intelligence needs to move from mere pattern recognition—no matter how sophisticated or fast—to causal explanation, which we covered in chapter 8. Many of the advances in artificial intelligence have come about through playing games, which requires a great deal of supervised trial-and-error learning with feedback about the correct answer. To beat a player at Go, for example—even one who's new to the game—a deep neural network must first observe millions of moves by expert players and play millions of practice games. Facebook's deep convolutional network needs thousands of examples to judge how a tower of just several toy blocks will fall, which is something a child would know intuitively.

In game-playing terms, artificial intelligence is still chess-like, trained to optimize the long-term reward of a particular action in a particular situation. It struggles to interpret novel input, such as reading a new style of handwriting. It cannot easily generalize or combine simple elements into complex concepts with infinite possibility—a feature of human thought and language known as *compositionality*. To make conversation, a neural network predicts the next sentence based on the previous one. In this sense it recalls the 1960s' MIT "Eliza" program, which faked its way through a conversation in "phrases tacked together like the sections of a prefabricated henhouse," as George Orwell once described the language of politicians. A half century later, the neural network shows more originality. "What is immoral?" Google researchers asked their neural conversational machine. "The fact that you have a child," it replied, somewhat ominously. To be fair, it and other intelligent personal assistants like Amazon's Alexa are designed to answer customer service questions or sell products, not to make original conversation compositionally or develop causal explanations about the world.

Compared to present artificial intelligence—distant future versions may be reading this and "laughing"—humans learn a lot more from much less. Learning both individually and socially, children can isolate variables and test causal hypotheses. Children who are taught *how* to learn, such as at Montessori schools, acquire measurable advantages in language, math, creativity, social interaction, and understanding. Humans can generalize explanatory concepts from just a few examples. To get closer to human creativity and flexibility of reasoning, artificial intelligence must become compositional and thus be able to generalize rather than simply look up each answer from an encyclopedic reference set.

This is precisely the goal of researchers who are experimenting with stochastic programs that can parse objects and goals into their essential components and then recombine them into new concepts and larger goals. This brings us back to the importance of memory. A breakthrough for DeepMind's neural computer came about through the integration of external read-write memory with the powerful neural network. This allows the computer to represent and manipulate complex data structures, and like a neural network, learn from the data. Just as humans have better working memory than chimpanzees, memory may be the key to humanlike artificial intelligence. Science fiction knows this. At the end of *2001: A Space Odyssey*, Hal becomes less human as his memory is unplugged. In HBO's *Westworld*, robots acquire human reasoning through their growing retention of personal memories.

As memory and artificial intelligence are integrated, however, and searchable digital records bring everything to one plane, we need to remember how to forget, as Elvis sang in 1955. Our metaphorical orca is not a fish but rather a mammal, and it occasionally comes up for air and to clear its mind. At the population scale, forgetting re-sorts existing variation, cleans the slate, and starts a new phylogenetic branch. We may owe our cultural modernity to this. About seventy-five thousand years ago, the volcanic eruption of Toba, in Sumatra, blanketing southern Asia in ash, might have left fewer than ten thousand people on the planet. Some paleoanthropologists believe the Upper Paleolithic era emerged out of Toba's ashes, with art, modern behaviors, and new technologies, all of which form the cultural foundation of humanity.

This is a pertinent question as memory and artificial intelligence become more integrated—literally, as machines begin to make decisions. Neuroadaptive technology already exists that can learn to interpret simple human intentions directly from brain activity.

While a person moves a cursor on the screen of a computer that is simultaneously doing real-time analysis of the person's brain activity at five hundred hertz, through dozens of electrodes configured around the scalp, the neuroadaptive system learns through trial and error how to translate the brain activity directly into the intended movement of the cursor. In short, the computer literally reads the person's mind. As with artificial intelligence in general, however, the question is, How big is the gap—in this case, between simple intentions, such as moving a cursor, and real thought and causal explanation?

ALONG COME MICE

In closing, we're reminded that evolution's direction is just too unpredictable to predict its future. As Walt Disney said in 1954 on *What Is Disneyland*, "I only hope we never lose sight of one thing: It was all started by a mouse." Steve Jobs might have said the same thing, except that Xerox had tried it before him. Remember the Xerox 8010? We didn't think so. The point is, no one could have guessed the trajectories that entertainment and the computer industry would take after those mice came along. They were products of countless cartoon characters and technological devices that came before them, and which one changed everything is only clear in hindsight.

One thing *can* be predicted with certainty, though: it's impossible to know exactly what cultural game changer is on the horizon. The best you can do is try to survey the pool of variation, but that's a tall order—an impossible one, actually. Our suggestion, not surprising, is to start with technology. For an entry point, you might browse the annual Consumer Electronics Show (now CES), which bills itself as "the launch pad for new innovation and technology that has changed the world." Just a few years ago, CES was awash with gadgets that connect to the Internet (and each other)—the "Internet of Things"—but in 2017, the show was, as *Forbes* noted, all "about making more things that create and use intelligence."

This leads us to a final question: Will technological and cultural evolution continue accelerating indefinitely, or is there a terminal velocity—a point of resistance that causes a "deceleration"? Already in 2017, for example, Facebook and European governments plan measures to curtail the spread of fake news, and surely artificial intelligence will be used—a case where selection is being bumped up in its balance with variation and transmission. As we wonder whether people one day will marry their intelligent, loving artificially intelligent assistant, like in the movie *Her*, we can also think about general evolutionary processes. To anticipate the future of cultural evolution, think about populations, not individuals, and certainly not yourself. How will variation, transmission, and selection be affected? What feedbacks will arise or be eliminated? Rather than latch onto a single prediction about what the future of culture holds, think like a Bayesian: How does this change the landscape of probabilities? What wave should we be surfing now, and how will we find the next wave after that? Willy and his orca friends call this fun, and you should, too.

BIBLIOGRAPHY

Chapter 1

Aiello, Leslie C., and Robin I. M. Dunbar. "Neocortex Size, Group Size, and the Evolution of Language." *Current Anthropology* 34 (1993): 184–193.

Aiello, Leslie C., and Peter Wheeler. "The Expensive-Tissue Hypothesis: The Brain and the Digestive-System in Human and Primate Evolution." *Current Anthropology* 36 (1995): 199–221.

Almaatouq, Abdullah, Laura Radaelli, Alex Pentland, and Erez Shmueli. "Are You Your Friends' Friend? Poor Perception of Friendship Ties Limits the Ability to Promote Behavioral Change." *PLOS ONE* 11, no. 3 (2016): e0151588.

Apicella, Coren L., Frank W. Marlowe, James H. Fowler, and Nicholas A. Christakis. "Social Networks and Cooperation in Hunter-Gatherers." *Nature* 481 (2012): 497–501.

Couzin, Ian D., Christos C. Ioannou, Güven Demirel, Thilo Gross, Colin J. Torney, Andrew Hartnett, Larissa Conradt, et al. "Uninformed

Individuals Promote Democratic Consensus in Animal Groups." *Science* 332 (2011): 1578–1580.

Dunbar, Robin I. M. "Neocortex Size and a Constraint on Group Size in Primates." *Journal of Human Evolution* 22 (1992): 469–493.

Dunbar, Robin I. M., and Susanne Shultz. "Evolution in the Social Brain." *Science* 317 (2007): 1344–1347.

Henrich, Joseph, Steven J. Heine, and Ara Norenzayan. "The Weirdest People in the World?" *Behavioral and Brain Sciences* 33 (2010): 61–135.

Hill, Kim R., Brian M. Wood, Jacopo Baggio, A. Magdalena Hurtado, and Robert T. Boyd. "Hunter-Gatherer Inter-Band Interaction Rates: Implications for Cumulative Culture." *PLOS ONE* 9, no. 7 (2014): e102806.

Horst, Heather, and Daniel Miller. *The Cell Phone: An Anthropology of Communication.* Oxford: Berg, 2006.

Hrdy, Sarah B. *Mothers and Others: The Evolutionary Origins of Mutual Understanding.* Cambridge, MA: Belknap, 2009.

Jessen, Sarah, and Tobias Grossmann. "Unconscious Discrimination of Social Cues from Eye Whites in Infants." *Proceedings of the National Academy of Sciences of the United States of America* 111 (2014): 16208–16213.

Lewis, Kevin, Marco Gonzalez, and Jason Kaufman. "Social Selection and Peer Influence in an Online Social Network." *Proceedings of the National Academy of Sciences of the United States of America* 109 (2012): 68–72.

Marshall, Lorna. "The Kin Terminology System of the !Kung Bushmen." *Africa* 27 (1957): 1–25.

Mesoudi, Alex. "An Experimental Simulation of the 'Copy-Successful-Individuals' Cultural Learning Strategy: Adaptive Landscapes, Producer-Scrounger Dynamics, and Informational Access Costs." *Evolution and Human Behavior* 29 (2008): 350–363.

Mesoudi, Alex, Lei Chang, Keelin Murray, and Jing Lu Hui. "Higher Frequency of Social Learning in China Than in the West Shows Cultural Variation in the Dynamics of Cultural Evolution." *Proceedings: Biological Sciences* 282 (2015): 2014–2209.

Mesoudi, Alex, and Michael J. O'Brien. "The Cultural Transmission of Great Basin Projectile-Point Technology I: An Experimental Simulation." *American Antiquity* 73 (2008): 3–28.

Pennisi, Elizabeth. "Conquering by Copying." *Science* 328 (2010): 165–167.

Rendell, Luke, Robert Boyd, Daniel Cownden, Magnus Enquist, Kimmo Eriksson, Marcus W. Feldman, Laurel Fogarty, et al. "Why Copy Others? Insights from the Social Learning Strategies Tournament." *Science* 328 (2010): 208–213.

Stone, Pamela K. "Biocultural Perspectives on Maternal Mortality and Obstetrical Death from the Past to the Present." *American Journal of Physical Anthropology* 159 (2016): 150–171.

Talhelm, Thomas, Xuemin Zhang, Shigehiro Oishi, Shimin Chen, Dongyuan Duan, Xuezhao Lan, and Shinobu Kitayama. "Large-Scale Psychological Differences within China Explained by Rice versus Wheat Agriculture." *Science* 344 (2014): 603–608.

Uhls, Yalda T., Minas Michikyan, Jordan Morris, Debra Garcia, Gary W. Small, Eleni Zgourou, and Patricia M. Greenfield. "Five Days at Outdoor Education Camp without Screens Improves Preteen Skills with Nonverbal Emotion Cues." *Computers in Human Behavior* 39 (2014): 387–392.

Wrangham, Richard. *Catching Fire: How Cooking Made Us Human.* New York: Basic Books, 2009.

Chapter 2

Ackroyd, Peter. *Foundation: The History of England from Its Earliest Beginnings to the Tudors.* London: Macmillan, 2011.

Bentley, R. Alexander, Joachim Wahl, T. Douglas Price, and Tim C. Atkinson. "Isotopic Signatures and Hereditary Traits: Snapshot of a Neolithic Community in Germany." *Antiquity* 82 (2008): 290–304.

Curry, Andrew. "Archaeology: The Milk Revolution." *Nature* 500 (2013): 20–22.

Evans-Pritchard, E. E. *The Nuer: A Description of the Modes of Livelihood and Political Institutions of a Nilotic People.* Oxford: Oxford University Press, 1940.

Hodder, Ian, and Craig Cessford. "Daily Practice and Social Memory at Çatalhöyük." *American Antiquity* 69 (2004): 17–40.

Kenoyer, Jonathan M. *Ancient Cities of the Indus Valley Civilization.* New York: Oxford University Press, 1998.

Loyn, Henry R. "Kinship in Anglo-Saxon England." *Anglo-Saxon England* 3 (1974): 197–209.

Maschner, Herbert D. G., and Katherine L. Reedy-Maschner. "The Evolution of Warfare." In *The Edge of Reason? Science and Religion in Modern Society*, edited by Alex Bentley, 57–64. London: Continuum Press, 2008.

Meyer, Christian, Christian Lohr, Detlef Gronenborn, and Kurt W. Alt. "The Massacre Mass Grave of Schöneck-Kilianstädten Reveals New Insights into Collective Violence in Early Neolithic Central Europe." *Proceedings of the National Academy of Sciences of the United States of America* 112 (2015): 11217–11222.

Schwandner-Sievers, Stephanie. "Humiliation and Reconciliation in Northern Albania." In *Dynamics of Violence*, edited by Georg Elwert, Stephan Feuchtwang, and Dieter Neubert, 133–152. Berlin: Duncker and Humblot, 1999.

Tehrani, Jamshid J. "The Phylogeny of Little Red Riding Hood." *PLOS ONE* 8, no. 11 (2013): e78871.

Wareham, Andrew. "The Transformation of Kinship and the Family in Late Anglo-Saxon England." *Early Medieval Europe* 10 (2001): 375–399.

Chapter 3

Acerbi, Alberto, and Alex Mesoudi. "If We Are All Cultural Darwinians What's the Fuss About?" *Biology and Philosophy* 30 (2015): 481–503.

Adamic, Lada A., Thomas M. Lento, Eytan Adar, and Pauline C. Ng. "Information Evolution in Social Networks." In *Proceedings of the Ninth ACM International Conference on Web Search and Data Mining*, 473–482. New York: ACM, 2014.

Andersen, Søren H. "Tybrind Vig." *Journal of Danish Archaeology* 4 (1985): 52–69.

Banaji, Shakuntala. "Slippery Subjects: Gender, Meaning, and the Bollywood Audience." In *The Routledge Companion to Media and Gender*, edited by Cynthia Carter, Linda Steiner, and Lisa McLaughlin, 493–502. Abingdon, UK: Routledge, 2015.

Boyd, Robert, and Peter J. Richerson. *Culture and the Evolutionary Process*. Chicago: University of Chicago Press, 1985.

Clutton-Brock, Tim. "Breeding Together: Kin Selection and Mutualism in Cooperative Vertebrates." *Science* 296 (2002): 69–72.

Dalrymple, William. "Homer in India." *New Yorker*, November 20, 2006, 48–55.

Evans-Pritchard, E. E. *The Nuer: A Description of the Modes of Livelihood and Political Institutions of a Nilotic People*. Oxford: Oxford University Press, 1940.

Kahan, Dan M. "What Is the 'Science of Science Communication'?" *Journal of Science Communication* 14 (2015): 1–10.

Kirby, Simon, Tom Griffiths, and Kenny Smith. "Iterated Learning and the Evolution of Language." *Current Opinion in Neurobiology* 28 (2014): 108–114.

Mesoudi, Alex, and Andy Whiten. "The Multiple Uses of Cultural Transmission Experiments in Understanding Cultural Evolution." *Philosophical Transactions of the Royal Society B: Biological Sciences* 363 (2008): 3489–3501.

Pinker, Steven A. *The Stuff of Thought: Language as a Window into Human Nature.* New York: Viking, 2007.

Purzycki, Benjamin G., Coren Apicella, Quentin D. Atkinson, Emma Cohen, Rita A. McNamara, Aiyana K. Willard, Dimitris Xygalatas, et al. "Moralistic Gods, Supernatural Punishment, and the Expansion of Human Sociality." *Nature* 530 (2016): 327–330.

Senghas, Ann, Sotaro Kita, and Asli Özyürek. "Children Creating Core Properties of Language: Evidence from an Emerging Sign Language in Nicaragua." *Science* 305 (2004): 1779–1782.

Wiedemann, Diana, D. Michael Burt, Russell A. Hill, and Robert A. Barton. "Red Clothing Increases Perceived Dominance, Aggression, and Anger." *Biology Letters* 11 (2015): 20150166.

Chapter 4

Corbey, Raymond, Adam Jagich, Krist Vaesen, and Mark Collard. "The Acheulean Handaxe: More Like a Bird's Song Than a Beatles' Tune?" *Evolutionary Anthropology* 25 (2016): 6–19.

Dawkins, Richard. *The Extended Phenotype: The Long Reach of the Gene.* Oxford: Oxford University Press, 1989.

Graça da Silva, Sara, and Jamshid J. Tehrani. "Comparative Phylogenetic Analyses Uncover the Ancient Roots of Indo-European Folktales." *Royal Society Open Science* (2016). doi:10.1098/rsos.150645.

Gray, Russell D., and Quentin D. Atkinson. "Language-Tree Divergence Times Support the Anatolian Theory of Indo-European Origin." *Nature* 426 (2003): 435–439.

Kitching, Ian J., Peter L. Forey, Christopher J. Humphries, and David M. Williams. *Cladistics: The Theory and Practice of Parsimony Analysis.* Oxford: Oxford University Press, 1992.

O'Brien, Michael J., and R. Lee Lyman. *Applying Evolutionary Archaeology: A Systematic Approach.* New York: Kluwer Academic, 2000.

O'Brien, Michael J., and R. Lee Lyman. *Cladistics and Archaeology.* Salt Lake City: University of Utah Press, 2003.

Smallwood, Ashley M., and Thomas A. Jennings, eds. *Clovis: On the Edge of a New Understanding.* College Station: Texas A&M University Press, 2015.

Tehrani, Jamshid J. "The Phylogeny of Little Red Riding Hood." *PLOS ONE* 8, no. 11 (2013): e78871.

Valverde, Sergi. "Review Article: Major Transitions in Information Technology." *Philosophical Transactions of the Royal Society B: Biological Sciences* 371 (2016): 20150450.

Valverde, Sergi, and Ricard V. Solé. "Punctuated Equilibrium in the Large-Scale Evolution of Programming Languages." *Journal of the Royal Society Interface* 12 (2015): 20150249.

Wilkins, Jayne, Benjamin J. Schoville, Kyle S. Brown, and Michael Chazan. "Evidence for Early Hafted Hunting Technology." *Science* 338 (2012): 942–946.

Chapter 5

Bentley, R. Alexander, Hallie R. Buckley, Matthew Spriggs, Stuart Bedford, Chris J. Ottley, Geoff M. Nowell, Colin G. Macpherson, et al. "Lapita Migrants in the Pacific's Oldest Cemetery: Isotopic Analysis at Teouma, Vanuatu." *American Antiquity* 72 (2007): 645–656.

Creanza, Nicole, Merritt Ruhlen, Trevor J. Pemberton, Noah A. Rosenberg, Marcus W. Feldman, and Sohini Ramachandran. "A Comparison of Worldwide Phonemic and Genetic Variation in Human Populations."

Proceedings of the National Academy of Sciences of the United States of America 112 (2015): 1265–1272.

Currie, Thomas E., Simon J. Greenhill, Russell D. Gray, Toshikazu Hasegawa, and Ruth Mace. "Rise and Fall of Political Complexity in Island South-East Asia and the Pacific." *Nature* 467 (2010): 801–804.

Fortunato, Laura. "Evolution of Marriage Systems." In *International Encyclopedia of the Social and Behavioral Sciences*, edited by James D. Wright, 14: 611–619. 2nd ed. Oxford: Elsevier, 2015.

Gibson, Mhairi A., and Eshetu Gurmu. "Land Inheritance Establishes Sibling Competition for Marriage and Reproduction in Rural Ethiopia." *Proceedings of the National Academy of Sciences of the United States of America* 108 (2011): 2200–2204.

Griffiths, Thomas L., Michael L. Kalish, and Stephan Lewandowsky. "Theoretical and Empirical Evidence for the Impact of Inductive Biases on Cultural Evolution." *Philosophical Transactions of the Royal Society B: Biological Sciences* 363 (2008): 3503–3514.

Grollemund, Rebecca, Simon Branford, Koen Bostoen, Andrew Meade, Chris Venditti, and Mark Pagel. "Bantu Expansion Shows That Habitat Alters the Route and Pace of Human Dispersals." *Proceedings of the National Academy of Sciences of the United States of America* 112 (2015): 13296–13301.

Holden, Clare J., and Ruth Mace. "'The Cow Is the Enemy of Matriliny': Using Phylogenetic Methods to Investigate Cultural Evolution in Africa." In *The Evolution of Cultural Diversity: A Phylogenetic Approach*, edited by Ruth Mace, Clare J. Holden, and Stephen Shennan, 217–234. London: University College Press, 2005.

Holden, Clare J., and Ruth Mace. "Spread of Cattle Led to the Loss of Matrilineal Descent in Africa: A Coevolutionary Analysis." *Proceedings of the Royal Society B: Biological Sciences* 270 (2003): 2425–2433.

Lewandowsky, Stephan, Thomas L. Griffiths, and Michael L. Kalish. "The Wisdom of Individuals: Exploring People's Knowledge about Everyday Events Using Iterated Learning." *Cognitive Science* 33 (2009): 969–998.

Murdock, George P. *Atlas of World Cultures.* Pittsburgh: University of Pittsburgh Press, 1981.

Vinyals, Oriol, and Quoc Le. "A Neural Conversational Model." ArXiv e-prints (2015): 1506.05869. https://arxiv.org/abs/1506.05869v3.

Watts, Joseph, Oliver Sheehan, Quentin D. Atkinson, Joseph Bulbulia, and Russell D. Gray. "Ritual Human Sacrifice Promoted and Sustained the Evolution of Stratified Societies." *Nature* 532 (2016): 228–231.

Chapter 6

Bentley, R. Alexander, Joachim Wahl, T. Douglas Price, and Tim C. Atkinson. "Isotopic Signatures and Hereditary Traits: Snapshot of a Neolithic Community in Germany." *Antiquity* 82 (2008): 290–304.

Corporation for National and Community Service. *Volunteer Growth in America: A Review of Trends since 1974.* Washington, DC, 2007.

Crowther, Alison, Leilani Lucas, Richard Helm, Mark Horton, Ceri Shipton, Henry T. Wright, Sarah Walshaw, et al. "Ancient Crops Provide First Archaeological Signature of the Westward Austronesian Expansion." *Proceedings of the National Academy of Sciences of the United States of America* 113 (2016): 6635–6640.

Giving Institute. *Giving USA: The Annual Report on Philanthropy for the Year 2015.* Chicago, 2016.

Haak, Wolfgang, Guido Brandt, Hylke N. de Jong, Christian Meyer, Robert Ganslmeier, Volker Heyd, Chris Hawkesworth, et al. "Ancient DNA, Strontium Isotopes, and Osteological Analyses Shed Light on Social and Kinship Organization of the Later Stone Age." *Proceedings of the National Academy of Sciences of the United States of America* 105 (2008): 18226–18231.

Hartocollis, Anemona. "College Students Protest, Alumni's Fondness Fades and Checks Shrink." *New York Times*, August 5, 2016, A1.

Horton, Mark, R. Alexander Bentley, and Philip Langton. "A History of Sugar: The Food Nobody Needs, but Everyone Craves."*Conversation*, October 30, 2015.

Hutcheson, John M. *Notes on the Sugar Industry of the United Kingdom*. Greenock, UK: McKelvie, 1901.

Jia, Sen, Thomas Lansdall-Welfare, and Nello Cristianini. "Time Series Analysis of Garment Distributions via Street Webcam." In *Image Analysis and Recognition*, edited by Aurélio Campilho and Fakhri Karray, 765–773. New York: Springer, 2016.

Lesthaeghe, Ron. "The Second Demographic Transition: A Concise Overview of Its Development." *Proceedings of the National Academy of Sciences of the United States of America* 111 (2014): 18112–18115.

National Center for Family Philanthropy. *Trends in Family Philanthropy: Examining the Present*. Washington, DC: Planning for the Future, 2015.

Pew Research Center. *Analysis of U.S. Decennial Census (1960–2000) and American Community Survey Data (2008, 2010–2014)*. Washington, DC: IPUMS, 2015.

Pollet, Thomas V., and Daniel Nettle. "Market Forces Affect Patterns of Polygyny in Uganda." *Proceedings of the National Academy of Sciences of the United States of America* 106 (2009): 2114–2117.

Robert Wood Johnson Foundation. *State of Obesity 2016*. Princeton, NJ, 2016.

Rosenblum, Nancy. *Good Neighbors: The Democracy of Everyday Life in America*. Princeton, NJ: Princeton University Press, 2016.

Ross, Cody T., Pontus Strimling, Karen P. Ericksen, Patrik Lindenfors, and Monique Borgerhoff Mulder. "The Origins and Maintenance of Female Genital Modification across Africa: Bayesian Phylogenetic Modeling of

Cultural Evolution under the Influence of Selection." *Human Nature* 27 (2016): 173–200.

Ruck, Damian, Daniel J. Lawson, and R. Alexander Bentley. Reduction in Religious Importance Precedes Economic Development. Forthcoming.

Steward, Miriam J., Edward Makwarimba, Linda I. Reutter, Gerry Veenstra, Dennis Raphael, and Rhonda Love. "Poverty, Sense of Belonging, and Experiences of Social Isolation." *Journal of Poverty* 13 (2009): 173–195.

Vance, J. D. *Hillbilly Elegy: A Memoir of a Family and Culture in Crisis.* New York: Harper Collins, 2016.

Chapter 7

Aral, Sinan, Lev Muchnik, and Arun Sundararajan. "Distinguishing Influence-Based Contagion from Homophily-Driven Diffusion in Dynamic Networks." *Proceedings of the National Academy of Sciences of the United States of America* 106 (2009): 21544–21549.

Bohannon, John. "Government 'Nudges' Prove Their Worth." *Science* 352 (2016): 1042.

Centola, Damon, and Andrea Baronchelli. "The Spontaneous Emergence of Conventions: An Experimental Study of Cultural Evolution." *Proceedings of the National Academy of Sciences of the United States of America* 112 (2015): 1989–1994.

Christakis, Nicholas A., and James H. Fowler. *Connected: The Surprising Power of Our Social Networks and How They Shape Our Lives.* New York: Little, Brown, 2009.

Dunbar, Robin I. M. "Neocortex Size and a Constraint on Group Size in Primates." *Journal of Human Evolution* 22 (1992): 469–493.

Foerster, Steffen, Mathias Franz, Carson M. Murray, Ian C. Gilby, Joseph T. Feldblum, Kara K. Walker, and Anne E. Pusey. "Chimpanzee Females

Queue but Males Compete for Social Status." *Scientific Reports* 6 (2016): 35404.

Hardin, Johanna, and Ghassan Sarkis, and the Pomona College Undergraduate Research Council. "Network Analysis with the Enron Email Corpus." *Journal of Statistics Education: An International Journal on the Teaching and Learning of Statistics* 23, no. 2 (2015): 1–27.

Henrich, Joe, and James Broesch. "On the Nature of Cultural Transmission Networks: Evidence from Fijian Villages for Adaptive Learning Biases." *Philosophical Transactions of the Royal Society B: Biological Sciences* 366 (2011): 1139–1148.

Hobaiter, Catherine, Timothée Poisot, Klaus Zuberbühler, William Hoppitt, and Thibaud Gruber. "Social Network Analysis Shows Direct Evidence for Social Transmission of Tool Use in Wild Chimpanzees." *PLOS Biology* 12, no. 9 (2014): e1001960.

Knappett, Carl, Ray Rivers, and Tim Evans. "The Theran Eruption and Minoan Palatial Collapse: New Interpretations Gained from Modeling the Maritime Network." *Antiquity* 85 (2011): 1008–1023.

Kovács, István, and Albert-László Barabási. "Destruction Perfected." *Nature* 524 (2015): 38–40.

Kristofferson, Kirk, Katherine White, and John Peloza. "The Nature of Slacktivism: How the Social Observability of an Initial Act of Token Support Affects Subsequent Prosocial Action." *Journal of Consumer Research* 40 (2014): 1149–1166.

Lieberman, Erez, Christoph Hauert, and Martin A. Nowak. "Evolutionary Dynamics on Graphs." *Nature* 433 (2005): 312–316.

Morone, Flaviano, and Hernán A. Makse. "Influence through Maximization in Complex Networks through Optimal Percolation." *Nature* 524 (2015): 65–68.

Sekara, Vedran, Arkadiusz Stopczynski, and Sune Lehmann. "Fundamental Structures of Dynamic Social Networks." *Proceedings of the National Academy of Sciences of the United States of America* 113 (2016): 9977–9982.

Silk, Joan B. "Social Components of Fitness in Primate Groups." *Science* 317 (2007): 1347–1351.

Watts, Duncan J. *Everything Is Obvious: Once You Know the Answer*. New York: Crown Business, 2011.

Chapter 8

Acerbi, Alberto, Vasileios Lampos, Philip Garnett, and R. Alexander Bentley. "The Expression of Emotion in 20th Century Books." *PLOS ONE* 8, no. 3 (2013): e59030.

Arthur, Brian. "Inductive Reasoning and Bounded Rationality." *American Economic Review* 84 (1994): 406–411.

Bollen, Johan, Huina Mao, and Xiaojun Zeng. "Twitter Mood Predicts the Stock Market." *Journal of Computational Science* 2 (2006): 1–8.

Bordino, Ilaria, Stefano Battiston, Guido Caldarelli, Matthieu Cristelli, Antti Ukkonen, and Ingmar Weber. "Web Search Queries Can Predict Stock Market Volumes?" *PLOS ONE* 7, no. 7 (2012): e40014.

Carter, Mike. "I Walked from Liverpool to London: Brexit Was No Surprise." *Guardian*, June 27, 2016.

Challet, Damien, and Ahmed Bel Hadj Ayed. "Predicting Financial Markets with Google Trends and Not So Random Keywords." *arXiv* (2014): 1307.4643.

Choi, Hyunyoung, and Hal Varian. "Predicting the Present with Google Trends." *Economic Record* 88 (2012): 2–9.

Christakis, Nicholas A., and James H. Fowler. *Connected: The Surprising Power of Our Social Networks and How They Shape Our Lives*. New York: Little, Brown, 2009.

Curme, Chester, Tobias Preis, H. Eugene Stanley, and Helen S. Moat. "Quantifying the Semantics of Search Behavior before Stock Market Moves." *Proceedings of the National Academy of Sciences of the United States of America* 111 (2014): 11600–11605.

Dong, Xianlei, and Johan Bollen. "Computational Models of Consumer Confidence from Large-scale Online Attention Data: Crowd-sourcing Econometrics." *PLOS ONE* 10, no. 3 (2015): e0120039.

Eichstaedt, Johannes C., Hansen A. Schwartz, Margaret L. Kern, Gregory Park, Darwin R. Labarthe, Raina M. Merchant, Sneha Jha, et al. "Psychological Language on Twitter Predicts County-Level Heart Disease Mortality." *Psychological Science* 26 (2015): 159–169.

Gayo-Avello, Daniel. "A Meta-analysis of State-of-the-art Electoral Prediction from Twitter Data." *Social Science Computer Review* 31 (2013): 649–679.

Jia, Sen, Thomas Lansdall-Welfare, and Nello Cristianini. "Time Series Analysis of Garment Distributions via Street Webcam." In *Image Analysis and Recognition*, edited by Aurélio Campilho and Fakhri Karray. 765–773. New York: Springer, 2016.

Li, Shirley. "The Wearable Device That Could Unlock a New Human Sense." *Atlantic*, April 14, 2015.

Lu, Zhixin, Kevin Klein-Cardeña, Steven Lee, Thomas M. Antonsen, Michelle Girvan, and Edward Ott. "Resynchronization of Circadian Oscillators and the East-West Asymmetry of Jet-lag." *Chaos* 26 (2016): 094811.

Mollan, Simon, and Ranald Michie. "The City of London as an International Commercial and Financial Center since 1900." *Enterprise and Society* 13 (2012): 538–587.

Orwell, George. "Politics and the English Language." *Horizon* 13 (1946): 252–265.

Preis, Tobias, Helen S. Moat, and H. Eugene Stanley. "Quantifying Trading Behavior in Financial Markets Using *Google Trends*." *Scientific Reports* 3 (2013): 1684.

Saavedra, Serguei, Jordi Duch, and Brian Uzzi. "Tracking Traders' Understanding of the Market Using E-Communication Data." *PLOS ONE* 6, 10 (2011): e26705.

Yelowitz, Aaron, and Matthew Wilson. "Characteristics of Bitcoin Users: An Analysis of Google Search Data." *Applied Economics Letters* 22 (2015): 1030–1036.

Chapter 9

Bendtsen, Kristian M., Florian Uekermann, and Jan O. Haerter. "Expert Game Experiment Predicts Emergence of Trust in Professional Communication Networks." *Proceedings of the National Academy of Sciences of the United States of America* 113 (2016): 12099–12104.

Bettencourt, Luís M. A., José Lobo, Dirk Helbing, Christian Kühnert, and Geoffrey B. West. "Growth, Innovation, Scaling, and the Pace of Life in Cities." *Proceedings of the National Academy of Sciences of the United States of America* 104 (2007): 7301–7306.

Bhattacharjee, Yudhijit. "The Mind of a Con Man." *New York Times Magazine*, April 28, 2013, MM44.

Bromham, Lindell, Xia Hua, Thomas G. Fitzpatrick, and Simon J. Greenhill. "Rate of Language Evolution Is Affected by Population Size." *Proceedings of the National Academy of Sciences of the United States of America* 112 (2015): 2097–2102.

Brown, William M., Lee Cronk, Keith Grochow, Amy Jacobson, C. Karen Liu, Zoran Popović, and Robert Trivers. "Retraction: Dance Reveals Symmetry Especially in Young Men." *Nature* 504 (2005): 470.

Butler, Declan. "Investigating Journals: The Dark Side of Publishing." *Nature* 495 (2013): 433–435.

Castelvecchi, Davide. "Physics Paper Sets Record with More Than 5,000 Authors." *Nature* (2015). doi:10.1038/nature.2015.17567.

Chawla, Dalmeet S. "Self-Citation Rates Higher for Men." *Nature* 535 (2016): 212.

Cross, Tim. "Technology Quarterly: After Moore's Law." *Economist*, March 12, 2016.

Derex, Maxime, Beugin Marie-Pauline, Bernard Godelle, and Michel Raymond. "Experimental Evidence for the Influence of Group Size on Cultural Complexity." *Nature* 503 (2013): 389–391.

de Sola Price, Derek J. "Networks of Scientific Papers." *Science* 149 (1965): 510–515.

Guo, Jeff. "I Have Found a New Way to Watch TV, and It Changes Everything." *Washington Post*, June 22, 2016.

Henrich, Joseph. "Demography and Cultural Evolution: Why Adaptive Cultural Processes Produced Maladaptive Losses in Tasmania." *American Antiquity* 69 (2004): 197–214.

Henrich, Joseph, Robert Boyd, Maxime Derex, Michelle A. Kline, Alex Mesoudi, Michael Muthukrishna, Adam T. Powell, et al. "Understanding Cumulative Cultural Evolution." *Proceedings of the National Academy of Sciences of the United States of America* 113 (2016): E6724–E6725.

Hidalgo, César. *Why Information Grows: The Evolution of Order, from Atoms to Economies.* New York: Basic Books, 2015.

Malakoff, David. "Are More People Necessarily a Problem?" *Science* 333 (2013): 544–546.

Mayer-Schönberger, Viktor, and Kenneth Cukier. *Big Data: A Revolution That Will Transform How We Live, Work, and Think.* Boston: Eamon Dolan, 2013.

Milot, Emmanuel, Francine M. Mayer, Daniel H. Nussey, Mireille Boisvert, Fanie Pelletier, and Denis Réale. "Evidence for Evolution in Response to Natural Selection in a Contemporary Human Population." *Proceedings of the National Academy of Sciences of the United States of America* 108 (2011): 17040–17045.

Moore, Gordon E. "Cramming More Components onto Integrated Circuits." *Electronics* (Basel) 38 (1965): 114–117.

Mulligan, Christina, and Timothy B. Lee. "Scaling the Patent System." *NYU Annual Survey of American Law* 68 (2012): 289–318.

Muthukrishna, Michael, and Joseph Henrich. "Innovation in the Collective Brain." *Philosophical Transactions of the Royal Society B: Biological Sciences* 371 (2016). doi:10.1098/rstb.2015.0192.

Pan, Raj K., Alexander M. Petersen, Fabio Pammolli, and Santo Fortunato. "The Memory of Science: Inflation, Myopia, and the Knowledge Network." *arXiv* (2016): 1607.05606v1.

Powell, Adam, Stephen Shennan, and Mark G. Thomas. "Late Pleistocene Demography and the Appearance of Modern Human Behavior." *Science* 324 (2009): 1298–1301.

Rands, Chris M., Stephen Meader, Chris P. Ponting, and Lunter G. Gerton. "8.2% of the Human Genome Is Constrained: Variation in Rates of Turnover across Functional Element Classes in the Human Lineage." *PLOS Genetics* 10, no. 7 (2014): e1004525.

Read, Dwight, and Claes Andersson. "The Evolution of Cultural Complexity: Not by the Treadmill Alone." *Current Anthropology* 57 (2016): 261–286.

Rumsey, Abby Smith. *When We Are No More: How Digital Memory Is Shaping Our Future.* New York: Bloomsbury Press, 2016.

Schläpfer, Markus, Luís M. A. Bettencourt, Sébastian Grauwin, Mathias Raschke, Rob Claxton, Zbigniew Smoreda, Geoffrey B. West, et al. "The Scaling of Human Interactions with City Size." *Journal of the Royal Society Interface* 11 (2014): 20130789.

Vaesen, Krist, Mark Collard, Richard Cosgrove, and Wil Roebroeks. "Population Size Does Not Explain Past Changes in Cultural Complexity." *Proceedings of the National Academy of Sciences of the United States of America* 113 (2016): E2241–E2247.

Wei, Pan, Gourab Ghoshal, Coco Krumme, Manuel Cebrian, and Alex Pentland. "Urban Characteristics Attributable to Density-Driven Tie Formation." *Nature Communications* 4 (2013): 1961.

Chapter 10

Albright, Jonathan. "The #Election2016 Micro-Propaganda Machine." *Medium*, November 18, 2016.

Atlantic Re:think. "Ask the Futurists: 10 Bold Predictions for 2030." *HPE Matter*, January 2017.

Baldwin, Matthew, and Joris Lammers. "Past-Focused Environmental Comparisons Promote Proenvironmental Outcomes for Conservatives." *Proceedings of the National Academy of Sciences of the United States of America* 114 (2017). doi: 10.1073/pnas.1610834113.

Bentley, R. Alexander, and Paul Ormerod. "Accelerated Innovation and Increased Spatial Diversity in U.S. Popular Culture." *Advances in Complex Systems* 15 (2012): 1150011.

Block, Barbara A., Heidi Dewar, Susanna B. Blackwell, Thomas D. Williams, Eric D. Prince, Charles J. Farwell, Andre Boustany, et al. "Migratory Movements, Depth Preferences, and Thermal Biology of Atlantic Bluefin Tuna." *Science* 293 (2001): 1310–1314.

Cadwalladr, Carole. "Google, Democracy, and the Truth about Internet Search." *Guardian*, December 4, 2016.

Chiasson, Dan. "All the Songs Are Now Yours." *New York Review of Books*, June 9, 2016.

Couzin, Ian D., Christos C. Ioannou, Güven Demirel, Thilo Gross, Colin J. Torney, Andrew Hartnett, Larissa Conradt, et al. "Uninformed Individuals Promote Democratic Consensus in Animal Groups." *Science* 332 (2011): 1578–1580.

Duck, Geraint, Goran Nenadic, Michele Filannino, Andy Brass, and David L. Robertson. "A Survey of Bioinformatics Database and Software Usage through Mining the Literature." *PLOS ONE* 11 (2016): e0157989.

Friedman, Thomas. *Thank You for Being Late: An Optimist's Guide to Thriving in the Age of Accelerations.* New York: Farrar, Straus and Giroux, 2016.

Graves, Alex, Greg Wayne, Malcolm Reynolds, Tim Harley, Ivo Danihelka, Agnieszka Grabska-Barwińska, Sergio Gómez Colmenarejo, et al. "Hybrid Computing Using a Neural Network with Dynamic External Memory." *Nature* 538 (2016): 471–476.

Henrich, Joseph, Robert Boyd, Maxime Derex, Michelle A. Kline, Alex Mesoudi, Michael Muthukrishna, Adam T. Powell, et al. "Understanding Cumulative Cultural Evolution." *Proceedings of the National Academy of Sciences of the United States of America* 113 (2016): E6724–E6725.

Lake, Brendan M., Tomer D. Ullman, Joshua B. Tenenbaum, and Samuel J. Gershman. "Building Machines That Learn and Think Like People." *Behavioral and Brain Sciences* 40 (2017).

Lerer, Adam, Sam Gross, and Rob Fergus. "Learning Physical Intuition of Block Towers by Example." *arXiv* (2016): 1603.01312.

Lillard, Angeline, and Nicole Else-Quest. "Evaluating Montessori Education." *Science* 313 (2006): 1893–1894.

Lo, Adeline, Herman Chernoff, Tian Zheng, and Shaw-Hwa Lo. "Framework for Making Better Predictions by Directly Estimating Variables' Predictivity." *Proceedings of the National Academy of Sciences of the United States of America* (2017). doi: 10.1073/pnas.1616647113.

MacCallum, Robert M., Matthias Mauch, Austin Burt, and Armand M. Leroi. "Evolution of Music by Public Choice." *Proceedings of the National Academy of Sciences of the United States of America* 109 (2012): 12081–12086.

McDermott, Josh H., Alan F. Schultz, Eduardo A. Undurraga, and Ricardo A. Godoy. "Indifference to Dissonance in Native Amazonians Reveals Cultural Variation in Music Perception." *Nature* 535 (2016): 547–550.

McGregor, Jim. "CES (Consumer Electronics Show) 2017 Preview." *Forbes*, January 3, 2017.

Mnih, Volodymyr, Koray Kavukcuoglu, David Silver, Andrei A. Rusu, Joel Veness, Marc G. Bellemare, Alex Graves, et al. "Human-Level Control through Deep Reinforcement Learning." *Nature* 518 (2015): 529–533.

Muchnik, Lev, Sinan Aral, and Sean J. Taylor. "Social Influence Bias: A Randomized Experiment." *Science* 341 (2013): 647–651.

Orwell, George. "Politics and the English Language." *Horizon* 13 (1946): 252–265.

Rohampton, Jimmy. "5 Social Media Trends That Will Dominate 2017." *Forbes*, January 3, 2017.

Rzeszutek, Tom, Patrick E. Savage, and Steven Brown. "The Structure of Cross-cultural Musical Diversity." *Proceedings of the Royal Society B: Biological Sciences* 279 (2012): 1606–1612.

Silver, David, Aja Huang, Chris J. Maddison, Arthur Guez, Laurent Sifre, George van den Driessche, Julian Schrittwieser, et al. "Mastering the Game of Go with Deep Neural Networks and Tree Search." *Nature* 529 (2016): 484–489.

Sridhar, Shrihari, and Raji Srinivasan. "Social Influence Effects in Online Product Ratings." *Journal of Marketing* 76 (2012): 70–88.

Strandburg-Peshkin, Ariana, Damien R. Farine, Ian D. Couzin, and Margaret C. Crofoot. "Shared Decision-making Drives Collective Movement in Wild Baboons." *Science* 348 (2015): 1358–1361.

Turk, Victoria. "Google's DeepMind Agrees New Deal to Share NHS Patient Data." *New Scientist*, November 22, 2016.

Vale, Gill L., Emma G. Flynn, Lydia Pender, Elizabeth Price, Andrew Whiten, Susan P. Lambeth, Steven J. Schapiro, et al. "Robust Retention and Transfer of Tool Construction Techniques in Chimpanzees (Pan Troglodytes)." *Journal of Comparative Psychology* 130 (2016): 24–35.

Wray, Christopher M., and Steven R. Bishop. "A Financial Market Model Incorporating Herd Behaviour." *PLOS ONE* 11 (2016): e0151790.

Wu, Chengkun, Jean-Marc Schwartz, Georg Brabant, Shao-Liang Peng, and Goran Nenadic. "Constructing a Molecular Interaction Network for Thyroid Cancer via Large-scale Text Mining of Gene and Pathway Events." Supplement 6, *BMC Systems Biology* 9 (2015): S5.

Wynn, Thomas, and Frederick L. Coolidge. "Beyond Symbolism and Language." *Current Anthropology* 51 (2010): S5–S16.

Yang, Shihao, Mauricio Santillana, and S. C. Kou. "Accurate Estimation of Influenza Epidemics Using Google Search Data via ARGO." *Proceedings of the National Academy of Sciences of the United States of America* 112 (2015): 14473–14478.

Zander, Thorsten O., Laurens R. Krol, Niels P. Birbaumer, and Klaus Gramann. "Neuroadaptive Technology Enables Implicit Cursor Control Based on Medial Prefrontal Cortex Activity." *Proceedings of the National Academy of Sciences of the United States of America* 113 (2016): 14898–14903.

INDEX

INDEX

Printed in the United States
by Baker & Taylor Publisher Services